THE MARCH 5, 1987, ECUADOR EARTHQUAKES

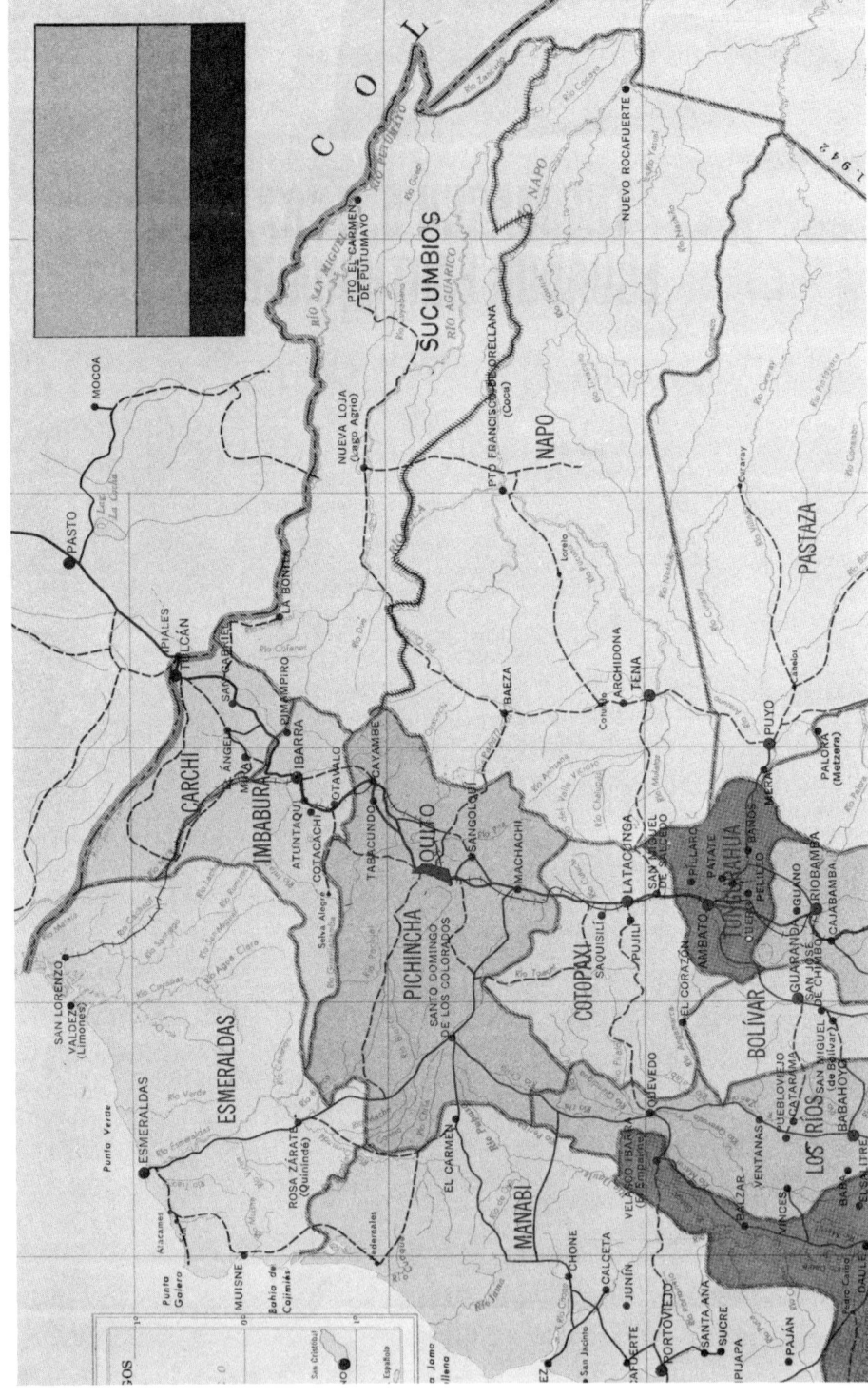

NATURAL DISASTER STUDIES

Volume Five

THE MARCH 5, 1987, ECUADOR EARTHQUAKES
MASS WASTING AND SOCIOECONOMIC EFFECTS

Study Team:

Robert L. Schuster (*Team Leader and Technical Editor*), Branch of Geologic Risk Assessment, U.S. Geological Survey, Denver, Colorado

Patricia A. Bolton, Battelle Institute, Seattle, Washington

Louise K. Comfort, Graduate School of Public and International Affairs, University of Pittsburgh, Pennsylvania

Esteban Crespo, School of Civil and Environmental Engineering, Cornell University, Ithaca, New York

Alberto Nieto, Department of Geology, University of Illinois, Urbana

Kenneth J. Nyman, School of Civil and Environmental Engineering, Cornell University, Ithaca, New York

Thomas O'Rourke, School of Civil and Environmental Engineering, Cornell University, Ithaca, New York

Contributing Authors:

José Egred, Instituto Geofísico, Escuela Politécnica Nacional, Quito, Ecuador

Alvaro F. Espinosa, Branch of Geologic Risk Assessment, U.S. Geological Survey, Denver, Colorado

Manuel García-Lopez, Departamento de Ingeniería Civil, Universidad Nacional de Colombia, Bogotá

Minard L. Hall, Instituto Geofísico, Escuela Politécnica Nacional, Quito, Ecuador

Galo Plaza-Nieto, Departamento de Geotécnica, Escuela Politécnica Nacional, Quito, Ecuador

Hugo Yepes, Instituto Geofísico, Escuela Politécnica Nacional, Quito, Ecuador

For:

Committee on Natural Disasters
Division of Natural Hazard Mitigation
Commission on Engineering and Technical Systems
National Research Council

NATIONAL ACADEMY PRESS
Washington, D.C. 1991

NOTICE: The project that is the subject of this report was approved by the Governing Board of the National Research Council, whose members are drawn from the councils of the National Academy of Sciences, the National Academy of Engineering, and the Institute of Medicine. The members of the committee responsible for the report were chosen for their special competences and with regard for appropriate balance.

This report has been reviewed by a group other than the authors according to procedures approved by a Report Review Committee consisting of members of the National Academy of Sciences, the National Academy of Engineering, and the Institute of Medicine.

The National Academy of Sciences is a private, nonprofit, self-perpetuating society of distinguished scholars engaged in scientific and engineering research, dedicated to the furtherance of science and technology and to their use for the general welfare. Upon the authority of the charter granted to it by the Congress in 1863, the Academy has a mandate that requires it to advise the federal government on scientific and technical matters. Dr. Frank Press is president of the National Academy of Sciences.

The National Academy of Engineering was established in 1964, under the charter of the National Academy of Sciences, as a parallel organization of outstanding engineers. It is autonomous in its administration and in the selection of its members, sharing with the National Academy of Sciences the responsibility for advising the federal government. The National Academy of Engineering also sponsors engineering programs aimed at meeting national needs, encourages education and research, and recognizes the superior achievements of engineers. Dr. Robert M. White is president of the National Academy of Engineering.

The Institute of Medicine was established in 1970 by the National Academy of Sciences to secure the services of eminent members of appropriate professions in the examination of policy matters pertaining to the health of the public. The Institute acts under the responsibility given to the National Academy of Sciences by its congressional charter to be an adviser to the federal government and, upon its own initiative, to identify issues of medical care, research, and education. Dr. Stuart Bondurant is acting president of the Institute of Medicine.

The National Research Council was organized by the National Academy of Sciences in 1916 to associate the broad community of science and technology with the Academy's purposes of furthering knowledge and advising the federal government. Functioning in accordance with general policies determined by the Academy, the Council has become the principal operating agency of both the National Academy of Sciences and the National Academy of Engineering in providing services to the government, the public, and the scientific and engineering communities. The Council is administered jointly by both Academies and the Institute of Medicine. Dr. Frank Press and Dr. Robert M. White are chairman and vice-chairman, respectively, of the National Research Council.

Library of Congress Catalog Card Number 91-67348
International Standard Book Number 0-309-04444-8

A limited number of copies of this monograph are available from:
Committee on Natural Disasters, HA 258
National Research Council
2101 Constitution Avenue, N.W.
Washington, DC 20418
202/334-3312

Additional copies are available for sale from:
National Academy Press
2101 Constitution Avenue, N.W.
Washington, DC 20418
202/334-3313
1-800-624-6242

Printed in the United States of America
S-314

NATURAL DISASTER STUDIES
An Investigative Series of the Committee on Natural Disasters

The Committee on Natural Disasters and its predecessors, dating back to the committee that studied the 1964 Alaska Earthquake, have conducted on-site studies and prepared reports reflecting their findings and recommendations on the mitigation of natural disaster effects. Objectives of the committee are to:

— record time-sensitive information immediately following disasters;
— provide guidance on how engineering and the social sciences can best be applied to the improvement of public safety;
— recommend research needed to advance the state of the art in the area of natural disaster reduction; and
— conduct special studies to address long-term issues in natural disasters, particularly issues of a multiple-hazard nature.

EDITOR
Riley M. Chung
National Research Council

EDITORIAL BOARD

Dennis S. Mileti, *Chair*
Colorado State University
Fort Collins

Norbert S. Baer
New York University
New York, New York

Earl J. Baker
Florida State University
Tallahassee

Arthur N. L. Chiu
University of Hawaii at Manoa
Honolulu

Hanna J. Cortner
University of Arizona
Tucson

Peter Gergely
Cornell University
Ithaca, New York

Joseph H. Golden
National Oceanic and
Atmospheric Administration
Washington, D.C.

Wilfred D. Iwan
California Institute
of Technology
Pasadena

Ahsan Kareem
University of Notre Dame
Notre Dame, Indiana

Dale C. Perry
Texas A&M University
College Station

William J. Petak
University of Southern
California
Los Angeles

Robert L. Schuster
U.S. Geological Survey
Denver, Colorado

SPONSORING AGENCIES
Federal Emergency Management Agency
National Oceanic and Atmospheric Administration
National Science Foundation

INVITATION FOR DISCUSSION

Materials presented in *Natural Disaster Studies* often contain observations and statements that inspire debate. Readers interested in contributing to the discussion surrounding any topic contained in the journal may do so in the form of a letter to the editor.

COMMITTEE ON NATURAL DISASTERS (1987–1990)

NORBERT S. BAER, Conservation Center of the Institute of Fine Arts, New York University, New York
EARL J. BAKER, Department of Geography, Florida State University, Tallahassee
ARTHUR N. L. CHIU, Department of Civil Engineering, University of Hawaii, Manoa
HANNA J. CORTNER, Water Resources Research Center, University of Arizona, Tucson
JOHN A. DRACUP, Civil Engineering Department, University of California, Los Angeles
DANNY L. FREAD, National Weather Service, National Oceanic and Atmospheric Administration, Silver Spring, Maryland
PETER GERGELY, Department of Structural Engineering, Cornell University, Ithaca, New York
JOSEPH H. GOLDEN, Chief Scientist Office, National Oceanic and Atmospheric Administration, Washington, D.C.
WILFRED D. IWAN, California Institute of Technology, Pasadena
AHSAN KAREEM, Civil Engineering Department, University of Notre Dame, Notre Dame, Indiana
DENNIS S. MILETI, Department of Sociology, Colorado State University, Fort Collins
JOSEPH PENZIEN, Department of Civil Engineering, University of California, Berkeley
DALE C. PERRY, Department of Construction Science, College of Architecture, Texas A&M University, College Station
WILLIAM J. PETAK, Institute of Safety and Systems Management, University of Southern California, Los Angeles
ROBERT L. SCHUSTER, Branch of Geologic Risk Assessment, U.S. Geological Survey, Denver, Colorado
RANDALL G. UPDIKE, Office of Earthquakes, Volcanoes and Engineering, U.S. Geological Survey, Reston, Virginia

Staff

RILEY M. CHUNG, Committee Director
EDWARD LIPP, Editor
SUSAN R. MCCUTCHEN, Administrative Assistant
SHIRLEY J. WHITLEY, Project Assistant

Liaison Representatives

WILLIAM A. ANDERSON, Earthquake Systems Integration, Division of Biological and Critical Systems, National Science Foundation, Washington, D.C.

BRUCE A. BAUGHMAN, Hazard Mitigation Branch, Public Assistance Division, Federal Emergency Management Agency, Washington, D.C.

FRED COLE, Office of U.S. Foreign Disaster Assistance, Agency for International Development, U.S. Department of State, Washington, D.C.

TERRY FELDMAN, Disaster Assistance Program, Federal Emergency Management Agency, Washington, D.C.

ROBERT D. GALE (deceased), U.S. Department of Agriculture/Forest Service, Washington, D.C.

EDWARD M. GROSS, Constituent Affairs and Industrial Meteorology Staff, National Weather Service, National Oceanic and Atmospheric Administration, Silver Spring, Maryland

WILLIAM HOOKE, Chief Scientist Office, National Oceanic and Atmospheric Administration, Washington, D.C.

PAUL KRUMPE, Office of U.S. Foreign Disaster Assistance, Agency for International Development, U.S. Department of State, Washington, D.C.

ELEONORA SABADELL, Division of Biological and Critical Systems, National Science Foundation, Washington, D.C.

GERALD F. WIECZOREK, Office of Earthquakes, Volcanoes and Engineering, U.S. Geological Survey, Reston, Virginia

ARTHUR J. ZEIZEL, Office of Natural and Technological Hazards Programs, State and Local Programs and Support, Federal Emergency Management Agency, Washington, D.C.

LAWRENCE W. ZENSINGER (Alternate), Office of Disaster Assistance Programs, State and Local Programs and Support, Federal Emergency Management Agency, Washington, D.C.

Acknowledgments

The study team was organized by the NAS/NRC Committee on Natural Disasters. Funding for travel for field studies and working meetings by the team was provided by NAS/NRC. Salaries for the study participants were provided by the individual organizations with which team members and other authors are affiliated. Dr. Espinosa's travel was funded by the U.S. Agency for International Development.

Within Ecuador, assistance and cooperation were provided by numerous governmental agencies, companies, and academic institutions. Although specific acknowledgments are presented in following chapters for the contributions of individual Ecuadorian organizations, we would especially like to thank officials and personnel of the Ministerio de Energía and the Instituto Ecuatoriano de Minería (INEMIN) and the Institute of Governmental Studies, who helped us both technically and logistically; Corporación Estatal Petrolera Ecuatoriana (CEPE)/Texaco, Inc., which provided helicopter transportation and lodging in the field area; and Instituto Ecuatoriano de Electrificación (INECEL), which provided technical support. We also would like to acknowledge the valuable technical advice and support we received from faculty members of the Instituto Geofísico of the Escuela Politécnica Nacional in Quito and Professor O. Lara of the Escuela Politécnica del Litoral in Guayaquil, who assisted in the initial stage of field reconnaissance.

Preface

After the March 5, 1987, earthquakes in Ecuador, the National Research Council, in cooperation with several other institutions, organized a postdisaster reconnaissance team to visit the disaster sites. The team was multidisciplinary, reflecting expertise in seismology, geology, geotechnical engineering, lifeline engineering, sociology, and the political sciences, in order to study the physical, social, and economic impacts on the nation resulting from the earthquakes. The team's effort in conducting its field work was greatly enhanced by a number of experts from local organizations and institutions. The eight chapters presented in this report are the contributions of the official members of the National Research Council team and their Ecuadorian colleagues working hand in hand to document and analyze these earthquakes. The individual chapters are the independent contributions of one or more of the individuals involved in the study. Efforts have been made by the team leader and the report's editor to eliminate duplications as much as possible. Remaining duplications are retained to allow the presentations to stand alone so that readers will not be burdened with cross referencing.

Robert L. Schuster
Team Leader and Technical Editor

Contents

EXECUTIVE SUMMARY 1

1 INTRODUCTION (*R. L. Schuster*) 11
   References, 20

2 GENERAL GEOLOGY OF NORTHEASTERN ECUADOR
   (*A. S. Nieto*) 23
   The Sierra, 23
   The Oriente, 26
   References, 28

3 TECTONICS AND SEISMICITY (*A. F. Espinosa, M. L. Hall, H. Yepes*) 29
   Introduction, 29
   Tectonic Setting, 30
   Seismicity and Focal Mechanisms, 31
   Recommendations, 40
   References, 40

4 INTENSITY AND DAMAGE DISTRIBUTION (*A. F. Espinosa, J. Egred, M. García-Lopez, E. Crespo*) 42
   Introduction, 42
   Relationship of Building Damage to Construction Practices, 43
   Intensity Distribution, 46
   References, 50

5 MASS WASTING AND FLOODING (*A. S. Nieto, R. L. Schuster, G. Plaza-Nieto*) 51
   Introduction, 51
   Landslides, 51
   Evolution of Mass-Wasting Processes, 72

xi

Flooding of River Valleys in the Vicinity of Reventador Volcano, 73
References, 80

6 EFFECTS ON LIFELINES (*E. Crespo, T. D. O'Rourke, K. J. Nyman*) 83
General Observations, 83
Characteristics of the Trans-Ecuadorian and Poliducto Pipelines, 86
Pipeline Damage, 87
Salado Pump Station Damage, 93
Trans-Ecuadorian Highway from Baeza to Lago Agrio, 97
Economic Consequences, 98
Summary, 98
Acknowledgments, 99
Reference, 99

7 LOCAL-LEVEL ECONOMIC AND SOCIAL CONSEQUENCES (*P. A. Bolton*) 100
Introduction, 100
Immediate Consequences and Emergency-Response Issues, 101
The Emergency Period in the Oriente, 103
The Emergency Period in the Sierra, 106
Emerging Long-Term Impacts, 109
Recovery Programs and Impacts in the Oriente, 110
Recovery Programs and Impacts in the Sierra, 114
Summary, 119
References, 120

8 ORGANIZATIONAL INTERACTION IN RESPONSE AND RECOVERY (*L. K. Comfort*) 122
Introduction, 122
Assumptions, 123
Organizational Interdependence in the Consequences of Disaster, 124
Organizational Networks in Disaster Response and Recovery Operations, 134
Acknowledgments, 151
Notes, 151

APPENDIXES
A. Disaster-Management Organizations—General, 156
B. International Organizations Involved in the March 1987 Disaster Operations, 158
C. Ecuadorian Organizations Involved in the March 1987 Disaster Operations, 160
D. Most-Damaged Areas, 163

THE MARCH 5, 1987, ECUADOR EARTHQUAKES:

Mass Wasting and Socioeconomic Effects

Executive Summary

The two earthquakes of magnitude 6.1 and 6.9 on March 5, 1987, occurred along the eastern slopes of the Andes Mountains in northeastern Ecuador. The epicenters were located in Napo Province, approximately 100 km ENE of Quito and 25 km N of Reventador Volcano. Ground shaking caused a number of moderate structural damages, primarily in areas near the epicenters. The occurrence of the earthquakes in an area of steep slopes covered by unstable volcanic and residual soils with high water contents, caused by heavy rainfall immediately prior to the earthquake, resulted in massive slope failures of high fluidity. Comparatively speaking, the economic and social losses directly due to earthquake shaking were small compared with the effects of catastrophic earthquake-triggered mass wasting (i.e., slumps and slides, debris flows, and debris avalanches) and flooding in the area adjacent to Reventador Volcano. Rock and earth slides, debris avalanches, and debris and mud flows east of the Andes resulted in the destruction or local severing of nearly 70 km of the Trans-Ecuadorian oil pipeline and the only highway from Quito to Ecuador's eastern rain forests and oil fields. Nearly all of the estimated 1,000 deaths from the earthquakes were a consequence of mass wasting and flooding. Economic losses were estimated at $1.0 billion; the effects of widespread denudation on the agricultural and hydroelectric development of the region are difficult to evaluate, but undoubtedly were very large.

General Geology of Northeastern Ecuador The geology of northeastern Ecuador and present-day physical processes related to geology are greatly influenced by the tectonic mechanisms responsible for the development of the Andes Mountains. The Andes have created three geologic and geomorphic zones: (1) the coastal plains (Costa) to the west, (2) the central mountainous area—the Andes (Sierra), and (3) the eastern lowlands (Oriente).

Seismicity and Tectonics Ecuador is exposed continually to earthquakes and other geologic hazards. The earthquake potential has always been a threat to the inhabitants of Ecuador, and thus coexisting with earthquake activity has become part of the Ecuadorian culture. In the past 80 years several large earthquakes (interplate events) have occurred in Ecuador's subduction zone. The March 1987 earthquakes, which occurred in the Andes, were not located in an active subduction zone along the plate boundaries. These intraplate earthquakes are shallow (10 to 14 km) events. Studies of earthquake potential, using conditional probability estimates, have shown a 66 percent probability for a great earthquake ($M_s \geq 7.7$) to take place along the subduction zone in the recurrence period of 1989-1999. The conditional probability estimates were evaluated using the historical and instrumental seismicity catalogue of the region; however, the historical record is poorly known for this region. A number of recommendations are made in this report to meet the urgent need to better evaluate the earthquake hazards in Ecuador. Among others, they include compilation of a more extensive and detailed historical earthquake catalogue for interplate and intraplate earthquakes, compilation of a historical earthquake catalogue for events that are associated with and have their origin in volcanism, deployment of sensitive seismological instruments in the region for ascertaining the level of seismicity and studying the joint focal mechanisms, and development of a working model of the tectonic regime of the region.

Intensity and Damage Distribution The areas of maximum Modified Mercalli Intensity (MMI) are concentrated in the meizoseismal area, attaining an Io = IX. However, much of the damage in this area should be classed as intensities VII and VIII. The problem encountered in the process of evaluating the intensity is that of "inconsistencies" in the MMI. For example, large landslides, such as were abundant in the high mountainous region near and around Reventador Volcano, as well as in other unstable regions in the high Andes, suggest an intensity greater than IX. Another factor that indicates high intensities (>IX) is surface faulting, such as that in areas near the epicentral region. Still another factor that yields higher intensities (X), as given in the intensity scale, is landslides from river banks and steep slopes due in some cases to water-saturated soils, shifted sand and mud, and water splashed over banks. In the MMI scale, "bridges destroyed" implies an intensity of XI. Yet in other cases in this event, although a bridge was indeed destroyed (but by flooding), wooden structures nearby sustained no damage. Other factors that may enter into the intensity-distribution pattern are seismic amplification effects, topographic seismic-wave amplification, influence of surficial soil conditions, and depth of the water table. Making the intensity assessment even more complicated are the highly mixed construction practices in this part of the country and the small and scattered population settlements in the mountainous areas.

Mass Wasting and Flooding Mass wasting and flooding account for the largest amount of destruction and number of deaths induced by the March 5, 1987, earthquakes. The area around Reventador Volcano includes the greatest intensity of landsliding triggered by the earthquakes. The earthquake epicenters lie a few kilometers to the N and W of the Reventador area. In this area, rainfall occurs throughout the year, but increases in intensity from March to July. Significantly, anomalously high precipitation occurred in the area in January and particularly in February 1987. On February 3 and 20, the gaging station just upstream from San Rafael Falls on the Coca River registered flow rates of 2,600 and 3,400 m^3/sec, respectively, which were 8 to 12 times higher than the average flow of the Coca River. More than 90 percent of the observed landslides began as shallow slips or slides of residual soils and highly weathered rock on the uppermost parts of the slopes of the main valleys or on the slopes of the lower-order tributaries. Average thicknesses of these slips were from 1.5 to 2.0 m, with a thickness range of a few decimeters to 5 m. The failing masses either were transformed into debris avalanches and then into debris flows or, in some cases, were reworked almost immediately into debris flows with high fluidity. A large number of the landslide scars displayed unweathered bedrock, attesting to the shallowness of the residual soil mantle.

The increase in denudation near Reventador Volcano was caused not only by its nearness to the epicenters, but also by other factors, namely relief, moisture conditions, elevation, and soil composition. The areas of near total denudation on the SW slope of the ancient cone correspond to an area of deep and dense dissection by parallel gullies. Here almost all the surfaces have slopes greater than 35 to 40°. Near total denudation in areas of dissection by gullies also has occurred along the walls of the deep canyons of several nearby rivers. In contrast, the slope on the north side of the ancient cone, which is not deeply dissected, is far less affected by landslides, even though it is closer to the epicenters and is less than 2 km from the area of almost total denudation.

One of the more striking characteristics of the mass wasting caused by the earthquakes was the effectiveness of the transport of materials and the volume of the materials from the slopes of the lowest-order tributaries to the flood plains of major streams. Two factors may have contributed to this characteristic. The first is the nature of the soils involved in the slope failures, and the second is the general morphology of the Reventador area.

Interruption of flow of the Coca River was witnessed immediately following the earthquakes, which indicates a strong possibility of natural damming of the river and/or its tributaries as the result of the earthquakes. Short-lived damming occurred in two ways: (1) "hydraulic" damming, in which stream flow, highly charged with debris, was impeded in passing through narrow bedrock constrictions in the stream channels, and (2) blockage of streams by debris flows issuing into the main stream from its tributaries.

Effects on Lifelines Damage to lifelines was severe in areas near the earthquake epicenters. Specifically, there was major damage to the Trans-Ecuadorian (crude oil) and Poliducto (propane) pipelines, as well as the principal highway linking Quito and Lago Agrio, the main town of the oil-producing region of Ecuador. Damage to lifelines in other areas was relatively light. The damage to these pipelines was so severe and widespread that it had a devastating economic impact on the nation, with possible repercussions felt worldwide. Approximately 40 km of the 498-km-long Trans-Ecuadorian pipeline had to be reconstructed, making this the single largest pipeline failure in history.

The Trans-Ecuadorian pipeline, commissioned in 1972, is composed primarily of 660-mm-diameter line pipe and associated pump and pressure-reducing stations. The pipeline is the main crude-oil transportation facility in Ecuador, conveying virtually all the oil from the eastern oil fields to a marine terminal port near Esmeraldas on the Pacific Ocean. The Poliducto pipeline is composed of 150-mm-diameter line pipe and associated compressor, pressure regulation, and pumping equipment. The line was built after the Trans-Ecuadorian pipeline was commissioned, and closely follows the right of way for the crude-oil line. The line is a multiproduct facility. It conveys different types of hydrocarbons at different times, including propane gas. It extends from Lago Agrio in the eastern oil field to Quito. It was destroyed and damaged at the same locations as the Trans-Ecuadorian pipeline, since the line was constructed following approximately the same route as the Trans-Ecuadorian pipeline.

The road parallel to the pipelines is the main transportation artery from Quito to the eastern oil field. Flooding destroyed the highway bridges on the Salado and Aguarico rivers, as well as large portions of the road between the Salado and Malo rivers. The Salado River bridge was replaced in 1988 by a Bailey bridge, which is still in use.

Loss of the Trans-Ecuadorian pipeline deprived Ecuador of 60 percent of its export revenue. As a consequence, the loss of this single lifeline had a dramatic effect on the country's economy. The total loss of revenue before the reconstructed line began its service in August 1987 was estimated at nearly $800 million. Added to that was the pipeline reconstruction cost of about $50 million.

The price of West Texas intermediate crude oil is often used as an index of the world price. News of the earthquakes and associated loss of the Trans-Ecuadorian pipeline was followed by a 6.25 percent increase in this index over the four trading days immediately following the earthquakes. Although oil prices had been climbing at the time of the earthquakes, market analysts claim that the news of Ecuador's suspension of oil exports encouraged trading at escalating prices. That is, the economic effects of the lifeline failure were not confined to a single country, but perhaps were felt on a worldwide basis by market speculation.

Local-Level Economic and Social Consequences The country's already deteriorating economy suffered a major blow when Ecuadorian oil production was disrupted by earthquake-related damage to the Trans-Ecuadorian pipeline. Since the oil fields had accounted for about 60 percent of the nation's export earnings, Ecuador's ability to meet its internal operating costs and to make interest payments on its foreign debt were severely impaired as the result of the pipeline failures. Within a week after the earthquakes, the national government instituted several extreme economic measures, including suspension of the external debt payment to private banks, increased fuel prices, a national austerity plan, and a price freeze on selected essential goods.

Both the emergency period after the earthquakes and the recovery process from the earthquakes were observed. One interesting point emerging from the observations is how people viewed their communities during the crisis. Aside from the original indigenous inhabitants, the populations in the towns and on the plantations of the Oriente were fairly recently arrived, and typically had left some other part of the country to seek economic opportunity as colonists to this area. These residents were not likely as yet to have long-term attachment to the land nor to have extensive and strong social ties to others in the area. One local storekeeper remarked that, right after the earthquake, the people had helped each other, but now things were "back to normal."

Another notable aspect of the earthquakes' effects on the Oriente was the impacts in Napo Province created by the loss of the Salado and Aguarico river bridges. First, the inaccessibility of the land along the approximately 67-km stretch of road between these two bridges prevented the return of the surviving farmers and plantation owners who had been evacuated from the area. Second, a large proportion of the 75,000 inhabitants of Napo Province were effectively cut off from the rest of Ecuador by the damage to the road between Baeza and Lago Agrio. Agricultural producers in the town of Lago Agrio and of the areas to the N and E in Napo Province suffered significant economic impacts as a result of not being able to transport their crops to market. It was estimated that the postearthquake production losses from abandonment of land or lack of access to markets amounted to about $7 million, based on the assumption that land access would be reestablished by the end of June, which it was not.

Another commonly observed issue during the recovery period of past natural disasters was again noted in this event, i.e., the dilemma of reconstructing communities in a timely manner versus taking the time and resources to provide safer housing and other facilities. On one hand, there is always concern that many families continue to live in houses that are even more vulnerable to future earthquakes than they were before the earthquakes. On the other hand, the reconstruction period also always provides one of the best opportunities for upgrading housing and making it safer, because interest in and awareness of earthquake threats are high.

Various interviews indicated that there was some controversy about what approach to reconstruction had the fewest drawbacks. Governmental officials claimed that the Indians, even though they are primarily farmers, had sufficient time to also work on the reconstruction of housing in their communities. They believe that the use of residents to provide construction labor makes the projects less costly and supposedly gives the residents a stronger sense of ownership for reconstruction activities. Others felt that the construction work was detrimental to the farming work and that this might lead to problems in the future if food production was inadequate. Another argument was given by some that it was not the best idea to have the villagers do the construction, since houses not properly designed and constructed will be likely to suffer damage during future earthquakes.

In the Sierra, special attention was given to the preservation of various damaged buildings that were felt to have historical and cultural value to the country. These buildings were considered important both for the citizens of Ecuador and for tourism. National funds were used for their repair.

A number of areas for further research are recommended from the assessment of economic and societal impacts on local communities as a result of the March 5, 1987, earthquakes. They include the extent to which technical assistance for construction, both in person and in the form of written materials, reaches the affected population, and what factors contribute to its effectiveness; the extent to which active efforts are made to distribute special instructional materials on handling the housing and health needs of disaster victims to persons at the local level who can assume responsibility for emergency programs; and the relationship between various types of recovery assistance and the promotion of local or national economic development, and policy development for this issue. Other recommendations are listed in Chapter 7.

Organizational Interaction in Response and Recovery The organizational interaction is particularly interesting in this event, given the multiple geographic locations of damage from the disaster, the multiple jurisdictional levels involved in disaster response and recovery activities, and the multiple perspectives required for timely and appropriate disaster assistance to the affected populations.

The earthquakes generated consequences of differing types and magnitudes in three geographic locations of Ecuador. The zone of primary impact, which included the epicenters of the earthquakes near Reventador Volcano, was located in western Napo Province. In this zone, major loss of life was caused by flash floods generated by massive landslides and debris flows. Major damage also occurred to the infrastructure systems. In the small towns affected by the disaster, the first response of the local organizations was directed toward meeting the human needs of the victims. The next major problems driving organizational interactions centered on destruction of the infrastructure. These problems included (1) reconstruction of the oil pipeline in geologically unstable territory, (2) loss of oil revenues and its consequent impact upon the national economy, (3) loss of the highway, bridges, and secondary roads for safe travel and economic activity of the resident population, and (4) reorientation and resettlement of local residents, severely shaken emotionally and economically by the disaster and struggling to cope with questions regarding an uncertain future in a zone of high seismic risk.

The zone of secondary impact from the disaster was the Sierra, where the principal problem was housing. Approximately 60,000 homes were damaged or rendered uninhabitable. Worse yet, what appeared to be a moderately severe event to other economic groups in this zone proved to be a disaster for those at the lowest economic level, whose homes were much more vulnerable to seismic risk and who had few resources for rebuilding. The third zone of impact from the disaster included the town of Lago Agrio and adjacent communities in eastern Napo Province. These communities suffered little structural damage and had no loss of life. The major problem generated by the earthquakes in this zone was isolation and economic deprivation, resulting from the destruction of the oil pipeline and the major route of land transportation. The cumulative effects of long-term isolation, unemployment, and lack of access to markets and supplies worsened with the prolonged period required for reconstruction of the infrastructure needed for the local economy, based upon oil production and agriculture.

Especially vulnerable were Indian communities along the Coca, Aguarico, Dué, Salado, and Papallacta rivers. Dependent upon the rivers for drinking water, food, and transportation, these communities suffered serious deprivation in the loss of these vital resources due to pollution and obstruction of the rivers. It was an interactive set of conditions that, unresolved, steadily worsened and overwhelmed the local resources of the residents and communities of this zone.

Differing consequences generated in three disaster zones required particular kinds of organizational actions for appropriate and timely response. As a result, the President of Ecuador established a national emergency committee, headed by the secretary of the National Security Council and an officer of the Ecuadorian Army, to direct the national disaster operation. However, the simultaneous needs of the populations in the three zones and the massive impact on the national economy from the combined loss of oil export revenues and cost of rebuilding the pipeline and transportation routes required resources beyond Ecuador's own capacity. As a result, some 22 nations responded to its needs. Yet the requirements for coordination and communication between participating nations and between the Ecuadorian levels of governmental jurisdiction in the simultaneous delivery of services to the three disaster zones elevated the complexity of organizational interaction.

The national emergency committee was also given the responsibility to coordinate activities by the international and voluntary organizations that participated in the disaster operation. In reality, the complexity of the operations and lack of communications facilities between national offices and the local governments necessitated that much of the actual work be done locally, with limited contribution from the national level.

As a result, voluntary charitable organizations played an important role in this disaster, particularly at the community level. These organizations, linked to the international community, provided resources that were not immediately available within Ecuador and initiated the design and implementation of disaster-assistance activities. The resulting pattern of organizational network performance, at times overlapping, at times operating independently, or uncoordinated, appears to be a function of at least four factors: (1) overall complexity of the disaster environment, (2) differing requirements of technology and resources for the problems addressed, (3) number and diversity of participating organizations, and (4) limited facilities, staff, and training for communication/coordination in disaster management.

Several needs were identified in this field study. They include the need (1) for improved communications, (2) for better coordination of action between the multiple organizations for improving performance in disaster operations, (3) to develop and disseminate more complete and systematic information for the wider set of organizations involved in disaster management, (4) to be especially sensitive when allocating disaster assistance to victims in communities with marginal economic standards, and (5) to evaluate performance in disaster operations and provide constructive feedback to participating organizations for improvement of their performance in future disasters.

The study also recommends several topics that warrant future research effort: (1) design and development of an interactive information system for decision support in disaster management, (2) design of interorganizational and interjurisdictional simulated postdisaster operations as a means of exploring the limits and capacities of human decision-making processes in disaster environments, (3) inquiry into the design and development of networks as appropriate organizational forms for the rapid mobilization, implementation, and evaluation of action in disaster management, and (4) inquiry into economies of resource management that will facilitate interorganizational participation.

1

Introduction

R. L. Schuster, U.S. Geological Survey, Denver, Colorado

On March 5, 1987, two earthquakes ($M_s=6.1$ at 2054 local time and $M_s=6.9$ at 2310 local time) occurred along the eastern slopes of the Andes Mountains in northeastern Ecuador. The epicenters were located in Napo Province (Figure 1.1), approximately 100 km ENE of Quito and 25 km N of Reventador Volcano (Figures 1.2, 1.3). Modified Mercalli Intensity (MMI) values as high as IX have been estimated for the epicentral area (Espinosa et al., this report). The shaking damaged structures in towns and villages near the epicentral area, particularly in the town of Ibarra (50 km NW of the epicenters), where two brick churches were severely damaged; several other brick buildings in Ibarra had to be reinforced subsequent to the quakes because of structural damage. In addition, considerable damage occurred to reinforced concrete buildings and foundations of wooden buildings in the village of Baeza (60 km SSW of the epicenters). In El Chaco village (50 km S of the epicenters), a steel-frame gymnasium, which was under construction, collapsed (Hakuno et al., 1988).

In spite of the seriousness of this structural damage, the economic and social losses directly due to earthquake shaking were small compared with the effects of catastrophic earthquake-triggered mass wasting and flooding in the area adjacent to Reventador Volcano (Figure 1.2). Rock and earth slides, debris avalanches, and debris and mud flows E of the Andes resulted in the destruction or local severing of nearly 70 km of the Trans-Ecuadorian oil pipeline and the only highway from Quito to Ecuador's eastern rain forests and oil fields. The total volume of earthquake-induced mass wasting has been estimated at from more than 75 million m^3 (Crespo et al., 1987) to about 110 million m^3 (Hakuno et al., 1988; Okusa et al., 1989). Economic losses have been estimated at $1 billion; the effects of widespread denudation on the agricultural and hydroelectric development of the

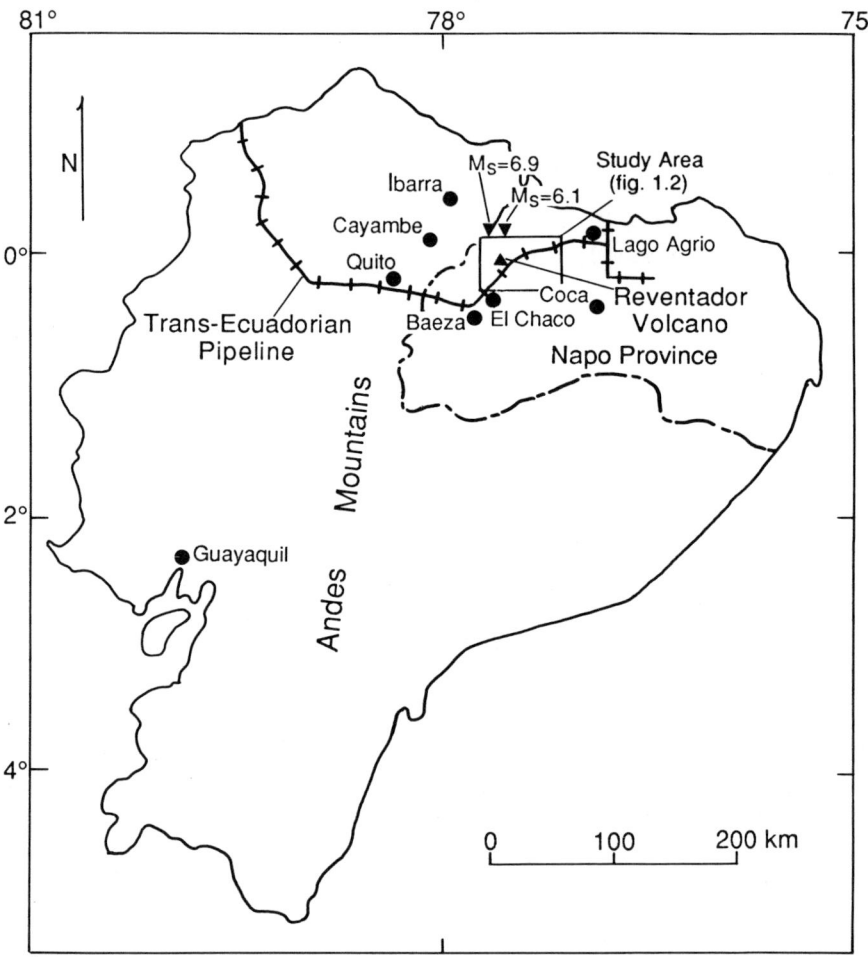

FIGURE 1.1 Index map of Ecuador showing locations of Napo Province, the Andes Mountains, Reventador Volcano (upright triangle), epicenters of the 1987 earthquakes (inverted triangles), the Trans-Ecuadorian oil pipeline, towns and villages (solid circles) that suffered structural damage from the earthquakes, and the mass-wasting study area (open rectangle; Figure 1.2).

region are difficult to evaluate, but undoubtedly were very large (Nieto and Schuster, 1988). Nearly all of the estimated 1,000 deaths from the earthquakes were a consequence of mass wasting and flooding. Because the mass wasting and flooding produced a high percentage of the economic and human losses resulting from these earthquakes, this report deals primarily with these processes, their socioeconomic effects, and the resulting social

INTRODUCTION

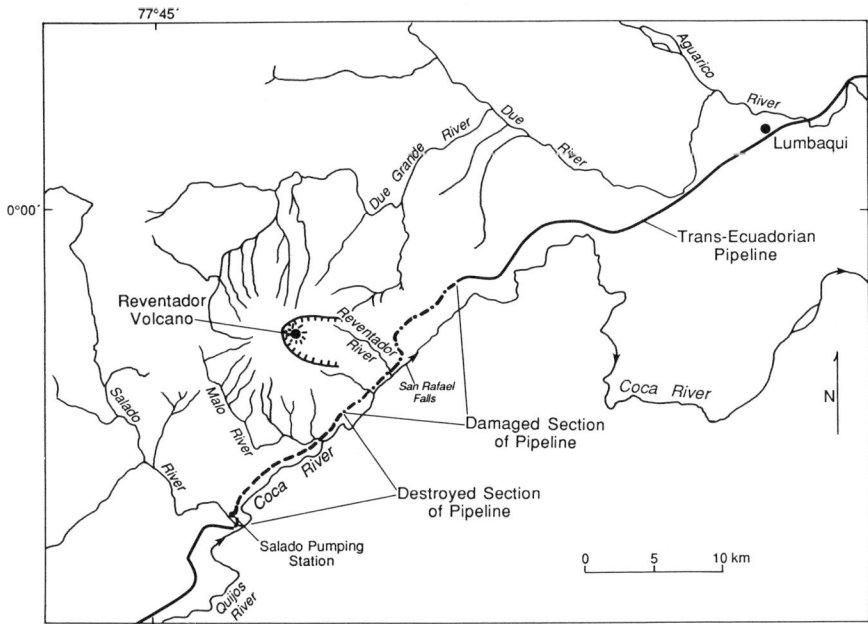

FIGURE 1.2 Area of study of mass wasting and flooding caused by the 1987 earthquakes, showing sections of damage to the Trans-Ecuadorian pipeline.

FIGURE 1.3 Reventador Volcano (elevation 3,562 m). (1978 photograph by S. D. Schwarz.)

implications. Observations related to these factors were made on site in Ecuador by the National Academy of Sciences/National Research Council (NAS/NRC) research team during the spring and summer of 1987.

Earthquakes are a major cause of mass wasting in many parts of the world. Earthquake-induced landslides have been documented for thousands of years; the earliest on record are landslides that dammed the Lo and Yi rivers in Hunan Province, China, in 1767 B.C. (Xue-Cai and An-ning, 1986). During the twentieth century, earthquake-induced landslides have caused tens of thousands of deaths and billions of dollars in economic losses (Keefer, 1984). In some cases, they have denuded thousands of square kilometers of unstable hillslopes. Particularly striking has been the denudation of jungle-covered, saturated slopes in tropical areas (Pain, 1972; Garwood et al., 1979).

A secondary hazard caused by earthquake-induced landslides is the formation of landslide dams. These natural stream blockages cause upstream flooding by stream impoundment, and they often breach catastrophically, causing major downstream flooding. Some of the world's most devastating floods have resulted from failure of large landslide dams that were formed by earthquake-induced landslides (Schuster and Costa, 1986; Costa and Schuster, 1988). In a review of more than 400 cases of historic landslide damming, Costa and Schuster (1991) have noted that about 35 percent of these blockages have been formed by landslides triggered by earthquakes.

The area of eastern Ecuador hardest hit by mass wasting due to the March 5, 1987, earthquakes was S of the epicentral area in the vicinity of Reventador Volcano (Figure 1.2). Most casualties from the earthquakes occurred in this region (Figure 1.4); the greatest damage to the Trans-Ecuadorian oil pipeline and highway occurred along the Coca River immediately SE of Reventador Volcano, upstream from beautiful San Rafael Falls (Figure 1.5). Because of the volcanic activity and river downcutting, the region exhibits strong relief. The average valley slopes range from 35 to 45°; before the 1987 landsliding, these slopes were generally covered by residual soils of variable thickness and by a dense, subtropical jungle.

The earthquake-induced slope failures were very fluid. About 600 mm of rain fell in the region in the month preceding the earthquakes; thus, the surface soils had a high moisture content. The slope failures commonly started as thin slips, which rapidly turned into very fluid debris avalanches and debris flows. The surficial materials and the thick jungle vegetation covering them flowed down the slopes into minor tributaries and then were carried into the major rivers (Salado, Quijos, Malo, Coca, Dué, Dué Grande, and Aguarico: Figure 1.2). Millions of tons of silty, gravelly sand, as well as tree remains and other organic matter, were deposited in the rivers (Figure 1.6). Many of the slopes were almost entirely denuded of their soil and jungle covers (Figure 1.7).

FIGURE 1.4 Cross erected as a memorial to nine Rodio S.A. employees who drowned in the Rio Coca near the mouth of the Rio Malo while trying to escape the flood caused by the March 5, 1987, earthquakes.

FIGURE 1.5 San Rafael Falls (height approximately 120 m) on the Coca River downstream from Reventador Volcano. The lip of the falls is formed by erosion-resistant lava flows. This falls is the approximate downstream limit of serious landslide/flood damage along the Coca River.

INTRODUCTION 17

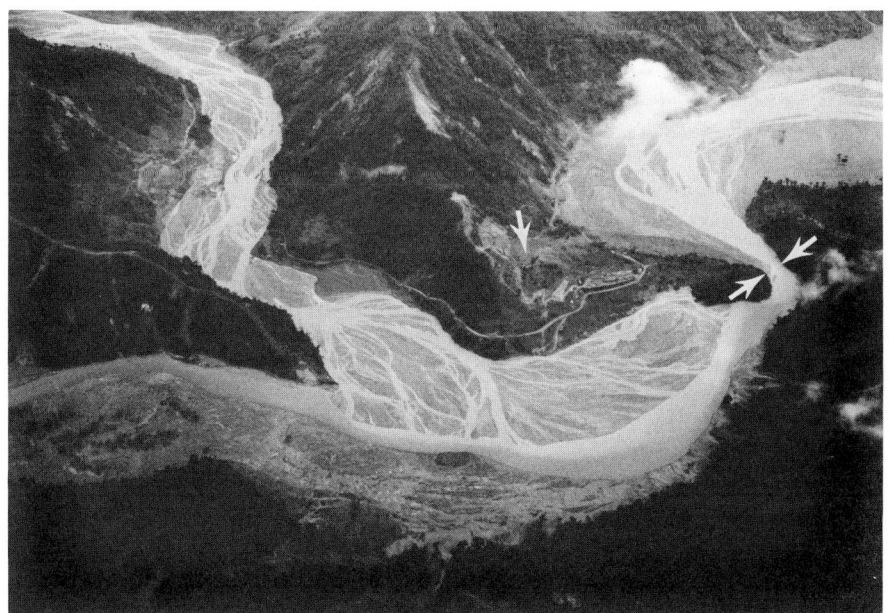

FIGURE 1.6 Aerial view of the confluence of the Quijos River (lower left) and the Salado River (upper left) to form the Coca River (flowing to the right). Postearthquake braided debris-flow and flood deposits are as much as 15 m thick in the valley bottoms. Bedrock constriction of the Coca River (indicated by two arrows near right edge of photo) probably caused short-lived damming of the river, which contributed to upstream flooding and rapid sedimentation. Note landslide (single arrow near center of photo) that badly damaged the Salado pumping station on the Trans-Ecuadorian oil pipeline.

Widespread stripping of saturated surficial materials and jungle cover from steep slopes by earthquake shaking similar to that which occurred in the Reventador area in 1987 has been noted in other humid tropical areas in a few similar catastrophes in this century. In September 1935, two shallow earthquakes (M=7.9 and M=7.0) in the Torricelli Range on the N coast of Papua New Guinea caused "hillsides to slide away, carrying with them millions of tons of earth and timber, revealing bare rocky ridges completely devoid of vegetation" (Marshall, 1937). Approximately 130 km^2 (8 percent of the region affected) was denuded by the landslides (Simonett, 1967; Garwood et al., 1979). Materials from the slides flooded the valleys, and, in some cases, blocked major rivers (Stanley, 1935). In November 1970, an M=7.9 earthquake, which was located along the N central coast of Papua New Guinea, triggered landslides that removed shallow soils and tropical forest vegetation from steep slopes in the Adelbert Range (Pain and Bowler, 1973). About 25 percent of the slope areas in the 240-km^2 area that was

FIGURE 1.7 NE (left) valley wall of the Malo River, showing extreme denudation of slopes due to slips/avalanches/flows caused by the March 5, 1987, earthquakes.

FIGURE 1.8 Aerial view of destruction of the Trans-Ecuadorian oil pipeline and adjacent highway by a debris flow issuing from a minor tributary of the Coca River. Location is near mouth of the Reventador River.

affected by landsliding were denuded (Pain, 1972). The soil debris and its cover of vegetation flowed off the slopes into drainage channels. Similarly, in 1976, two shallow earthquakes (M=6.7 and M=7.0) struck the sparsely populated SE coast of Panama, causing huge areas of landsliding. Garwood et al. (1979) calculated that the slides denuded approximately 54 km^2 (12 percent of the affected region of 450 km^2). Although the M=9.2 earthquake that struck southern Chile in May 1960 occurred in an area of temperate forest rather than in tropical jungle, it caused slope failures in the Valdivian Andes similar to those in Papua New Guinea and Panama. Veblen and Ashton (1978) estimated that more than 250 km^2 of forest slopes were denuded by mass wasting in the 1960 event.

Given the size of the mass-wasting catastrophe in the Reventador area, damage caused by direct impact of deep-seated slides or slumps was secondary to that caused by thin slips, avalanches, flows, and floods. Although individual slides did some damage to the Trans-Ecuadorian pipeline (Figure 1.8), roads, and structures, the greatest destruction of property was caused by flood surges in the main rivers (Figure 1.9). Because of antecedent precipitation, the rivers were near flood stage before the earthquakes occurred, so that the large volumes of landslide debris that flowed into the

FIGURE 1.9 Destruction of Trans-Ecuadorian oil pipeline and highway by flood erosion on the left bank of the Coca River as a result of the March 5, 1987, earthquakes.

valleys further raised the river stages (Nieto and Schuster, 1988). It is likely that the highest flood surges were caused by breaching of short-lived dams on tributaries carrying large sediment loads, by large debris flows moving directly off valley walls, or by debris blockages at narrow constrictions of the river channels (Figure 1.10).

In summary, interrelated multiple hazards produced the catastrophic events of March 5, 1987, in the Reventador area. The tragic occurrence of two large earthquakes within 3 hr in an area of heavy antecedent rainfall, and steep slopes covered by unstable volcanic and residual soils with high water contents, resulted in massive slope failures of high fluidity. The large volumes of these slope failures and the breaching of the resulting ephemeral debris dams caused the flood surges that were responsible for most of the damage.

FIGURE 1.10 Upstream view of Salado River, showing location of bedrock constriction (white arrows) that caused short-lived damming of the river. Note trimline in jungle cover along lower valley wall upstream of the river constriction. This trimline indicates position of the shoreline that was formed by damming of the river to a level 10-15 m above current river level.

REFERENCES

Costa, J. E., and R. L. Schuster. 1988. The formation and failure of natural dams. Geological Society of America Bulletin, 100:1054–1068.

Costa, J. E., and R. L. Schuster. 1991. Documented Historical Landslide Dams from Around the World. U.S. Geological Survey Open-File Report 91–239:486.

Crespo, E., K. J. Nyman, and T. D. O'Rourke. 1987. Ecuador Earthquakes of March 5, 1987. Earthquake Engineering Research Institute Special Earthquake Report, 4.

Garwood, N. C., D. P. Janos, and N. Brokaw. 1979. Earthquake-caused landslides: A major disturbance to tropical forests. Science 205(4410, 7 September):997–999.

Hakuno, M., S. Okusa, and M. Michiue. 1988. Study Report of Damage Done by the 1987 Earthquakes in Ecuador. Research Report on Unexpected Disasters, Research Field Group, Natural Disasters and the Ability of the Community to Resist Them, Supported by the Japanese Ministry of Education, Culture, and Science (Grant No. 626010221), 38.

Keefer, D. K. 1984. Landslides caused by earthquakes. Geological Society of America Bulletin 95:406–421.

Marshall, A. J. 1937. Northern New Guinea, 1936. Geographical Journal 89(6):489–506.
Nieto, A. S., and R. L. Schuster. 1988. Mass wasting and flooding induced by the 5 March 1987 Ecuador earthquakes. Landslide News, newsletter of The Japan Landslide Society, 2:1–3.
Okusa, S., M. Hakuno, and M. Michiue. 1989. Distribution of factors of safety for natural slope stability—an example of Ecuador. Proceedings of the Japan-China Symposium on Landslides and Debris Flows. Niigata, October 3, Tokyo, October 5, The Japan Landslide Society and The Japan Society of Erosion Control Engineering:273–278.
Pain, C. F. 1972. Characteristics and geomorphic effects of earthquake-initiated landslides in the Adelbert Range, Papua New Guinea. Engineering Geology 6(4):261–274.
Pain, C. F., and J. M. Bowler. 1973. Denudation following the November 1970 earthquake at Madang, Papua New Guinea. Zeitschrift für Geomorphologie Suppl. Bd. 18:92–104.
Schuster, R. L., and J. E. Costa. 1986. A perspective on landslide dams. Pp. 1–20 in Landslide Dams: Processes, Risk, and Mitigation, R. L. Schuster, ed. Geotechnical Special Publication No. 3, American Society of Civil Engineers.
Simonett, D. S. 1967. Landslide distribution and earthquakes in the Bewani and Torricelli Mountains, New Guinea. Pp. 64–84 in Landform Studies from Australia and New Guinea, J. N. Jennings and J. A. Mabbutt, eds. Canberra: Australian National University Press.
Stanley, G. A. V. 1935. Preliminary notes on the recent earthquake in New Guinea. Australian Geographer 2(8):8–15.
Veblen, T. T., and D. H. Ashton. 1978. Catastrophic influences on the vegetation of the Valdivian Andes, Chile. Vegetation 36(3):149–167.
Xue-Cai, F., and G. An-ning. 1986. The principal characteristics of earthquake landslides in China. Geología Applicata e Idrogeología 21(2):27–45.

2

General Geology of Northeastern Ecuador

A. S. Nieto, Department of Geology, University of Illinois, Urbana

The geology of northeastern Ecuador and present-day physical processes related to geology are greatly influenced by the tectonic mechanisms responsible for the development of the Andes Mountains. Both geology and active physical processes (landsliding, volcanism, erosion, weathering) are complex and varied. The reader is referred to classic works on these subjects (Tschopp, 1953; Lewis et al., 1956; Ham and Herrera, 1963; Feininger, 1975; Hall, 1977; Baldock, 1982a,b; Feininger, 1987; etc.). Oil and mineral exploration has provided the impetus for detailed studies on the geology of NE Ecuador. The following paragraphs draw liberally on the above-mentioned sources.

The Andes have created three geologic and geomorphic zones: (1) the coastal plains (Costa) to the west, (2) the central mountainous area—the Andes (Sierra) themselves, and (3) the eastern lowlands (Oriente) (Figures 2.1, 2.2). Figure 2.1 presents a geomorphic/geologic framework of Ecuador. The Costa is a region of low relief and low elevation W of the Cordillera Occidental (Western Cordillera), one of the two major branches of the Ecuadorian Andean Mountains. Much of the ground surface of the Costa consists of Quaternary volcanic and alluvial soils that may be unstable under earthquake loads. However, the energy of the March 5, 1987, earthquakes had dissipated to insignificant levels by the time it reached the Costa. Therefore, this region is not discussed here. Only the geology of the eastern two-thirds of Ecuador (the Sierra and the Oriente) is described in the remainder of this chapter, because this was the area most seriously affected by the earthquakes.

THE SIERRA

The Sierra is bounded on the W by a suture zone (Jubones Fault) that defines the eastern edge of the Costa and on the E by the back-arc fold-and-thrust belt of the Oriente Province (Figure 2.1). The Sierra traverses the

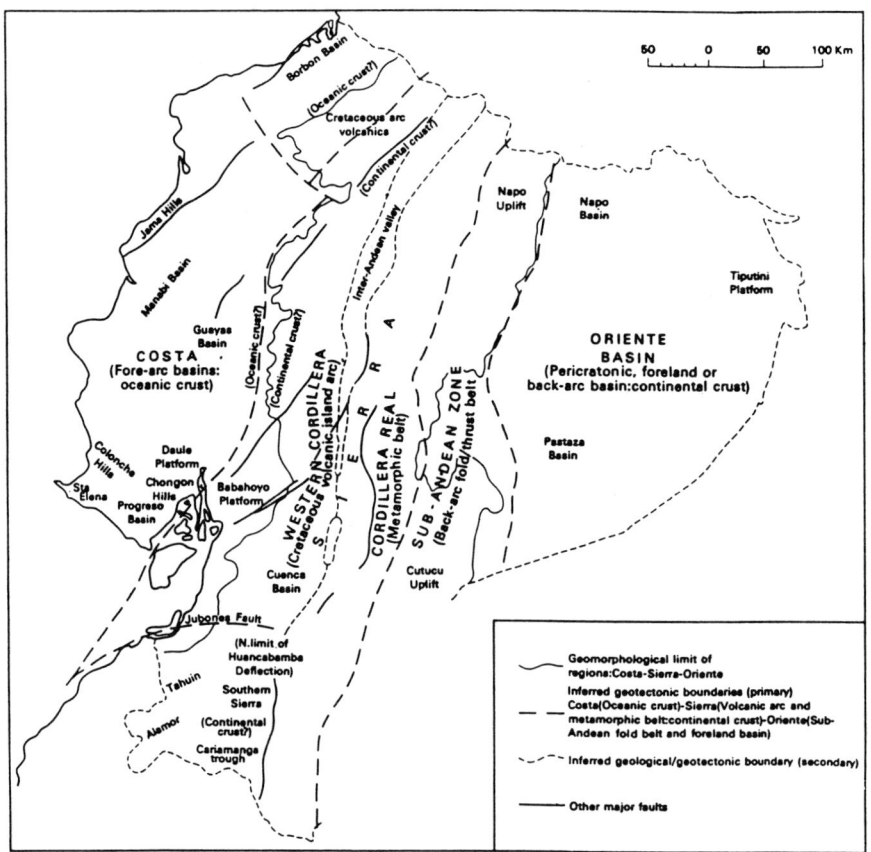

FIGURE 2.1 Geomorphic/geologic framework of Ecuador (after Baldock, 1982b).

length of the country and is only about 150 km wide, much narrower than the rest of the Andes. Three geologic and geomorphic zones exist within the Sierra: the Cordillera Occidental (Western Cordillera), the Inter-Andean Valley, and the Cordillera Real (Eastern Cordillera).

The origin of the Cordillera Occidental has been given at least two interpretations. Baldock (1982a) interpreted the zone as a sequence of volcanic-arc sediments (the Macuchi Formation), which were deposited in the Upper Cretaceous to Eocene and were tectonically emplaced at a later time. The basement is continental crust except in the extreme N. Feininger (1987) also interpreted the sediments as volcanic in origin. However, high Bouguer gravity anomalies throughout the Costa and the Cordillera Occidental led Feininger to interpret the entire area W of the Inter-Andean Valley to the N and the upper Amazon Basin to the S as allochthonous terrane underlain by oceanic crust.

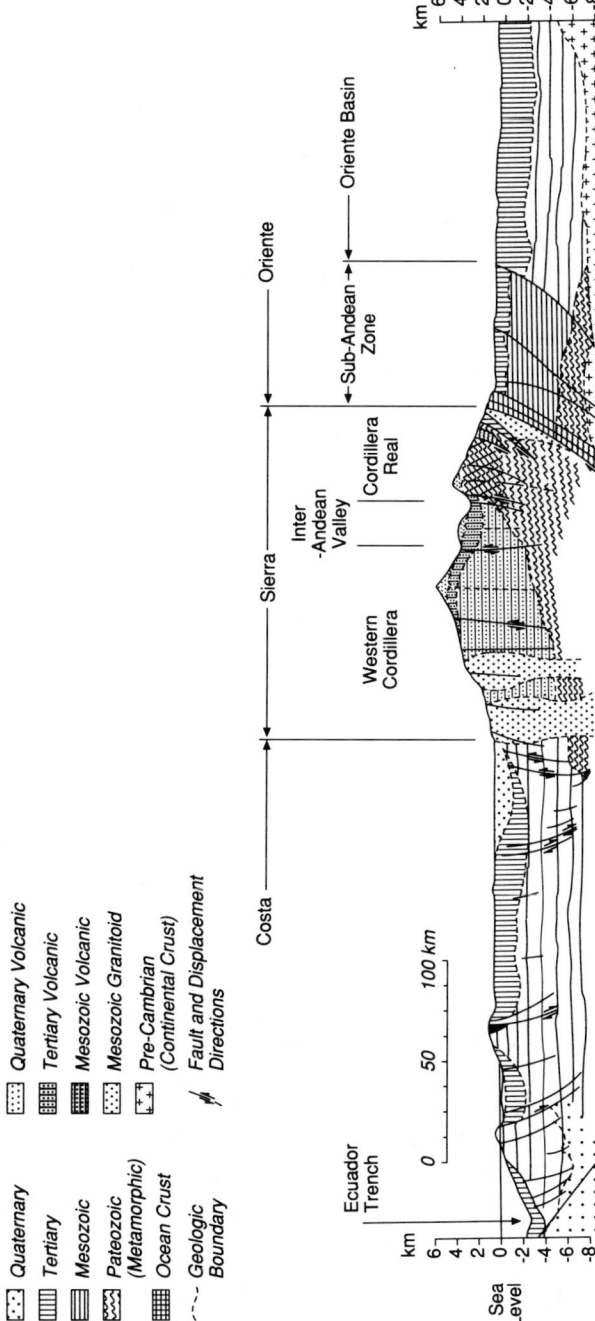

FIGURE 2.2 E-W geologic cross section through Ecuador at approximately 1° 30' S latitude (after Baldock, 1982b; simplified by Hakuno et al., 1988).

The Macuchi Formation consists of a thick sequence of pillow lavas and andesitic volcaniclastic deposits. It is overlain by Paleocene and Eocene volcaniclastic and marine sediments. In some areas, Miocene sediments overlie the Macuchi Formation. Small late Tertiary intrusive bodies intersect the Macuchi Formation. Neogene to Quaternary volcanic deposits obscure the Macuchi at higher elevations of this zone.

The Inter-Andean Valley is a graben situated between the two Cordilleras (Figure 2.1). Intermontane basins that occupy this region are more prevalent in the N and become smaller and less continuous to the S. These high valleys (2,500 to 3,000 m in elevation) are filled with Quaternary sedimentary and pyroclastic deposits. The most important of these volcanic deposits is volcanic ash known as "cangahua." This ash, of aeolian origin, is fine-grained, largely unstratified, and weakly cemented. At times it resembles loess, or slightly cemented, highly porous sandstone. The cangahua is prone to slope failure.

The Cordillera Real is bounded on the W by the Inter-Andean Valley and on the E by the Sub-Andean Zone. Paleozoic, and perhaps older, metamorphic rocks are dominant in this region. These metamorphic rocks were probably formed during a Caledonian orogenic event (Baldock, 1982a). Subsequent orogenic events, including the Laramide and Andean orogenies, likely affected the rocks of the Cordillera Real. Lithologies present in the region include a thick sequence of Paleozoic muscovite-biotite schists and a sequence of mica schists and chlorite schists (Llanganates Group). Isoclinal folding in these metamorphic rocks has been observed in only a few areas. Crenulation cleavage that overprints the folding may imply a subsequent orogenic event. The majority of fracturing is the result of effects of the Andean orogeny and of Neogene uplift. The region is sporadically covered by Quaternary volcanic rocks (lavas) and sediments (cangahua), which usually are unconformable with underlying metamorphic rocks.

THE ORIENTE

The Oriente (Figures 2.1, 2.2) consists of two distinct structural zones and physiographic provinces: the Oriente Basin and the Sub-Andean Zone. Physiographically, the Sub-Andean Zone consists of foothills rising to elevations of up to 2,000 m. East-flowing rivers have deeply dissected these foothills. The climate varies from tropical in the eastern portions to subtropical in the higher western reaches. Rainfall is high everywhere; as a consequence, rates of weathering are generally high. The Sub-Andean Zone, which borders the Cordillera Real (Figures 2.1, 2.2), is a back-arc fold-thrust belt tectonically associated with the Andes (Baldock, 1982a). Two folded features, the Napo uplift to the N and the Cutucu uplift to the S, are separated by the Lorocachi arch. Reventador Volcano is located on the Napo uplift.

The Oriente (or Amazon) Basin lies east of the Sub-Andean Zone. This is a gently warped basin that represents a more stable tectonic history than that of the Sub-Andean Zone. The stratigraphy of the two zones is similar.

The Guyana Shield (Precambrian crystalline rocks) makes up the basement of the Oriente Basin. In the early Paleozoic, part of the Oriente underwent transgression and sedimentation. The Caledonian orogeny affected the Oriente region only by shifting the axis of sedimentation eastward. Paleozoic lithologies include shales and quartzitic sandstones of the Devonian Pumbuiza Formation and limestones of the Carboniferous Macuma Formation. Three tectonic events during the Mesozoic and late Tertiary had little tectonic effect on the Oriente. Similarly, the Laramide orogeny of late Eocene and Oligocene had a minimal influence on the Oriente. Basin sedimentation resulted in deposition of fresh-water and terrestrial lithologies. Middle Jurassic to late Cretaceous redbeds (Chapiza Formation) and clastics and pyroclastics (Misahualli Member of the Chapiza Formation) underlie the Cretaceous Hollin-Napo-Tena Formations. In the Reventador area, the Mishualli Member increases in thickness and is predominantly volcanic. Rock types associated with the Hollin-Napo-Tena group include quartzitic sandstones (Hollin Formation), which are reservoir rocks for petroleum in NE Ecuador. The overlying Napo Formation consists of shales, limestones, and sandstones, all of marine origin. The sandstones may also be reservoir rocks. The Tena is composed of redbeds and shale. Redbeds and some sandstones and clays represent early Cenozoic deposition.

The major deformation of the Sub-Andean Zone took place in the late Miocene and Pliocene. Overthrusting and uplift during this time were responsible for the present segregation of the two zones of the Oriente. Major thrust faults associated with this event generally trend NNE to SSW (Figures 2.1 and 2.2). Two additional sets of faults or fractures have been observed in the Reventador Volcano area of the Napo uplift (INECEL, 1987). There is an indication of low-grade metamorphism associated with some of the major (NNE to SSW) thrust faults in the Reventador area of the Napo uplift.

Quaternary clastic sedimentation in the Oriente includes a variety of deposits, from lavas and pyroclastics of all grain sizes to colluvial/alluvial materials (piedmont fans) to alluvial fills.

Two major areas of Quaternary volcanism occur in the Sub-Andean Zone (Hall, 1977) (Figure 2.1). Sumaco Volcano (20 km SE of Baeza, Figure 1.1) deposited alkaline undersaturated basalts. Reventador Volcano (Figures 1.1, 1.3) exhibits a more typical suite of lithologies, including andesitic basalts and a significant amount of pyroclastic and lahar deposits. Both volcanos are situated on the Napo uplift and overlie Cretaceous rocks. While Sumaco and Reventador volcanoes are both considered active, Reventador alone has undergone historic and frequent volcanic activity. The morphological evolution of Reventador Volcano began in the Pliocene. The original volcanic cone (Paleo

Reventador I) collapsed initially (perhaps in the Pliocene); the remains of this collapse created the "Complejo Volcánico Basal" of INECEL (INECEL, 1987). Activity resumed in the Holocene, and culminated with the collapse of another cone—Paleo Reventador II—about 20,000 years ago; this second cone has been referred to as the "Paleo Reventador" by INECEL. This second collapse dammed the Coca River and caused the deposition of 20 m of lacustrine sediments that have been dated radioactively by INECEL. The material from the two collapses reached the right bank of the Coca, creating terranes of volcanic debris that are present there today.

REFERENCES

Baldock, J. W. 1982a. Geology of Ecuador—Explanatory Bulletin of the National Geological Map of the Republic of Ecuador. Ministerio de Recursos Naturales y Energéticos, Dirección General de Geología y Minas, Quito, 70.

Baldock, J. 1982b. National Geological Map of the Republic of Ecuador. Ministerio de Recursos Naturales y Energéticos, Dirección General de Geología y Minas, Quito, scale 1:1,000,000.

Feininger, T. 1975. Origin of petroleum in the Oriente of Ecuador. American Association of Petroleum Geologists Bulletin 59(7):1166–1175.

Feininger, T. 1987. Allochthonous terranes in the Oriente of Ecuador and northwestern Peru. Canadian Journal of Earth Sciences 24:266–278.

Hakuno, M., S. Okusa, and M. Michiue. 1988. Study Report of Damage Done by the 1987 Earthquakes in Ecuador. Research Field Group, Natural Disasters and the Ability of the Community to Resist Them, Supported by the Japanese Ministry of Education, Culture, and Science (Grant No. 62601022), Research Report on Unexpected Disasters No. B-62-2, 38.

Hall, M. L. 1977. El volcanismo en el Ecuador. Publicación del I.P.G.H., Sección Nacional del Ecuador, Quito, 73–80.

Ham, C. K., and L. J. Herrera, Jr. 1963. Role of the Subandean Fault System on tectonics of Eastern Peru and Ecuador. Pp. 47–61 in Backbone of the Americas—Tectonic History from Pole to Pole. O. E. Childs and B.W. Beebe, eds. American Association of Petroleum Geologists Memoir 2.

INECEL (Instituto Ecuatoriano de Electrificación). 1987. Proyecto hidroeléctrico Coca—Codo Sinclair—estudios de factibilidad. Ministerio de Energía y Minería, Republic of Ecuador, 24 plus figures.

Lewis, G. E., H. J. Tschopp, and J. G. Marks. 1956. Ecuador. Pp. 250–291 in Handbook of South American Geology—an Explanation of the Geologic Map of South America. W. F. Jenks, ed. Geological Society of America Memoir 65.

Tschopp, H. J. 1953. Oil explorations in the Oriente of Ecuador, 1938-1950. Bulletin of the American Association of Petroleum Geologists 37(10):2303–2357.

3

Tectonics and Seismicity

A. F. Espinosa, U.S. Geological Survey, Denver, Colorado
M. L. Hall, Escuela Politécnica Nacional, Quito, Ecuador
H. Yepes, Escuela Politécnica Nacional, Quito, Ecuador

INTRODUCTION

Ecuador is exposed continually to earthquakes and other geologic hazards. In particular, earthquake potential has always been a threat to the inhabitants of Ecuador, so that coexisting with earthquake activity has become part of the Ecuadorian culture.

From the hazards point of view, one must differentiate between earthquakes of tectonic origin and those associated with volcanism. In the past 80 years several large earthquakes (interplate events) have occurred in Ecuador's subduction zone, and their rupture mechanisms were varied (Kanamori and McNally, 1982). Shallow intraplate events, such as the March 1987 earthquakes, occur in the Andes, distant from the active subduction zone. These earthquakes created a serious socioeconomic problem for the country and triggered hundreds of associated geologic hazards—massive landslides, subsidence, liquefaction, impoundment of rivers, and other effects common to earthquakes that have occurred in similar geologic settings (Espinosa, 1979).

Although several destructive intraplate earthquakes have occurred, no systematic probabilistic studies have been done to ascertain the earthquake hazard. Using macroseismic information from chronicles of the sixteenth century (Egred, 1988), an approximate epicenter at 0.14°S and 78.27°W was assigned to an earthquake that took place in April 1541 with a magnitude of 7.0. On August 16, 1868, a great earthquake occurred at 0.31°N and 78.18°W with a magnitude of 7.7. On June 23, 1925, a magnitude 6.8 earthquake with a depth of focus of 180 km occurred and was located at 0.0°, 77°W. A great interplate earthquake with an M_w = 8.8 took place on January 31, 1906, with a length of rupture on the order of 500 km (Kanamori and McNally, 1982). Other large interplate earthquakes have occurred in northwestern Ecuador, including those of May 14, 1942 (M_s = 7.9) and January 19, 1958 (M_s = 7.8). The March 5, 1987,[1], main event (M_s = 6.9) occurred in interior Ecuador, in a highly faulted zone of Jurassic intrusive and Cretaceous metamorphic rocks. This earthquake and its largest foreshock (M_s = 6.1) are the subject of this chapter.

TECTONIC SETTING

Active seismicity occurs continuously for more than 6,000 km along the western edge of South America. The oceanic Nazca Plate is steadily being subducted eastward under the continent along a well-defined Wadati-Benioff zone. The most tectonically notable feature of the South American Plate is the Andean Mountains, which share a common tectonic pattern from Colombia in the N to southern Chile. The major physiographic features of the Andes are the result of the subduction of the Pacific lithosphere beneath the South American continent.

Three distinct tectonic regimes characterize the Nazca Plate oceanward of Colombia and Ecuador. Between latitudes 1 and 7°N, the ocean bottom physiography is nearly flat. Its age varies progressively from 10 to 26 million years toward the N (Lonsdale and Klitgard, 1978); its subduction to the E, under Colombia, coincides with a row of active stratovolcanoes. Between latitudes 2 and 4°S, the ocean bottom in front of the Ecuadorian Trench is a fractured and complex zone, 230 km wide. This region is cut by several oceanic fracture zones with NE-trending directions, identified as the Grijalva, Alvarado, and Sarmiento fractures. As this region is subducted under the South American continent, it may behave as a separate microplate independent of the adjacent plates (Pennington, 1981; Hall and Wood, 1985).

In between these two tectonic regimes, between latitude 1°N and 2°S, a submarine mountain range called the Carnegie Ridge, which was generated by the Galapagos mantle plume, collides against the South American continent. This mountain range is approximately 300 km wide and 3 km high, and rests upon older oceanic crust more than 16 million years old. During the past 25 million years, the Nazca Plate has moved eastward at a relative plate velocity of 5 cm/yr (Pilger, 1983), subducting the E-W- oriented Carnegie Ridge under central Ecuador. Lonsdale (1978) estimated that the subduction of the Carnegie Ridge started about 2 or 3 million years ago, while Pennington (1981) estimated an even earlier beginning. Probably its subduction began 5 to 6 million years ago, when, because of the

---

[1]Times and dates in this chapter are based on local Ecuadorian time except where specifically noted as being U.T.C. (Universal Time Coordinate), which is equivalent to Greenwich Mean Time.

difficulty in subducting this large physiographic feature, the Nazca Plate's eastward journey slowed, which finally permitted the uniform lava emission from the Galapagos plume to build subaerial volcanoes before the plate moved onward (J. W. Spence and M. L. Hall, personal communication, 1987).

Where the subduction of the Carnegie Ridge takes place, the trench is shallow, the coastal region is being uplifted, and extensive and chemically diverse volcanism occurs in the Andes. The mode of faulting and seismicity of the region may be related to the subduction of the Carnegie Ridge. Other tectonic features can also be attributed to this subduction, such as the greater height of the Ecuadorian Andes in this zone, the formation of stratovolcanoes, and active strike-slip faults (Hall and Wood, 1985). The Yaquina fracture zone (Lonsdale and Klitgard, 1978) is not parallel to adjacent N-S trending transform faults in the Panama Basin, but swings westward as it approaches the Carnegie Ridge, suggesting that the subduction of the Nazca Plate in this region is being slowed considerably, most probably because of the difficulty in subducting a very large physiographic feature such as the Carnegie Ridge (Hall and Wood, 1985). The collision of the Carnegie Ridge with continental Ecuador has altered the tectonic stress distribution along this convergent margin, resulting in the creation of numerous faults with NW-SE and NE-SW trends. Well-known fault systems oriented NE-SW include those of the Gulf of Guayaquil, La Pallatanga, and the Alausi-Guamote Valley faults, among others. Several of the destructive earthquakes that have occurred in Ecuador, the Riobamba in 1797 and the Alausi in 1961, among others, have been correlated with these NE-SW trending faults. Major lineaments and faults with NW-SE orientations have been identified by Hall and Wood (1985) as delimiting regions of tectonic segmentation, the most important ones being the Esmeraldas-Pastaza and the Rio Mira-Salado lineaments. The intersection of several sets of conjugate faults occurs in the Inter-Andean Valley, a region well known for its high seismicity and destructive earthquakes, such as the Ibarra earthquake of 1868, the Ambato earthquake of 1949, and the Pastocalle earthquakes of 1944 and 1976.

In the region of the March 5, 1987, earthquake, four principal sets of faults and lineaments have been clearly identified on satellite images (LANDSAT No. 010060; unpublished identification and interpretation by M. L. Hall, 1986) as shown in Figure 3.1. The Abra fault trends N40°E and passes through the western foot of Reventador Volcano (Figure 3.1). This fault has a surface expression of at least 180 km in length, as identified from the LANDSAT images. Parallel lineaments with a length of up to 80 km are visible about 10 km to the NW from the first fault. Another fault lying E of Reventador Volcano and trending parallel to the above faults is known as the Rio Quijos fault (Figure 3.1). These faults are all considered

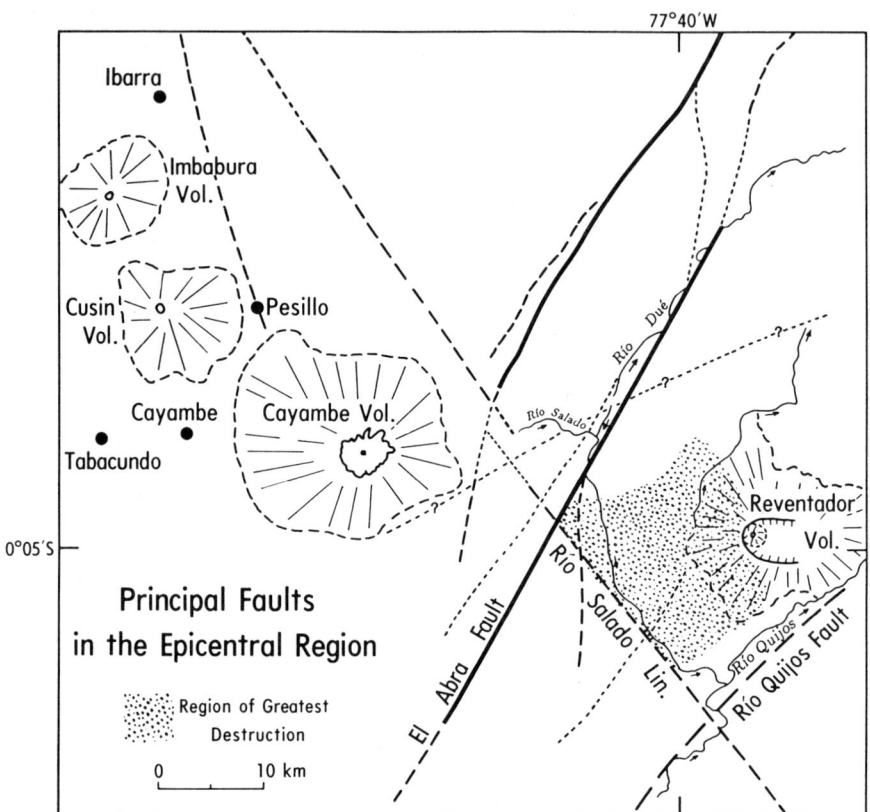

FIGURE 3.1 Map showing the fault systems around the epicentral region (LANDSAT imagery No. 010060, unpublished identification and interpretation by M. L. Hall, 1986).

to be steep reverse faults. The fourth system is associated with the Rio Mira-Salado lineament; this system trends N29°W and is 260 km long. Some of these tectonic features are also shown in Figure 3.2, along with the location of Reventador Volcano.

SEISMICITY AND FOCAL MECHANISMS

The shallow seismicity of the region from latitude 5°S to 5°N and longitude 75° to 85°W, for the time period from 1962 through 1987, is shown in Figure 3.3A. In this figure all the instrumentally recorded shallow earthquakes with uncertain depths are assigned depths of 33 km; these are likely to be shallow, as depth phases are not well separated. Earthquakes with body- and/or surface-wave magnitudes equal to or greater than 4.5 are plotted. The highest seismicity in this figure occurs in the Ecuador-Colombia zone from latitude 1 to 4°N and longitude 77 to 80°W, most of which are aftershocks following the December 12, 1979, ($M_s = 7.7$) earthquake. There is also a diffuse seismic zone in the interior of Ecuador that shows no definitive pattern that can be directly associated with the regional tectonic structures at this scale. The earthquake distributions for depths of focus between 33 and 100 km, and between 100 and 300 km, respectively, are shown in Figures 3.3B and 3.3C. The seismic activity portrayed in Figure 3.3B shows a concentration of maximum seismicity in the SW part of Ecuador. The lithosphere in the N part of the Nazca Plate has a gentle angle of subduction, and its geometrical configuration is not uniform, because of the presence of the subducting Carnegie Ridge. The highest seismicity at depths greater than 100 km is shown in Figure 3.3C and is concentrated between latitude 1 and 2°S and at approximately longitude 78°W. The damaging earthquakes of historic times have been large, shallow earthquakes whose epicenters were near urban areas.

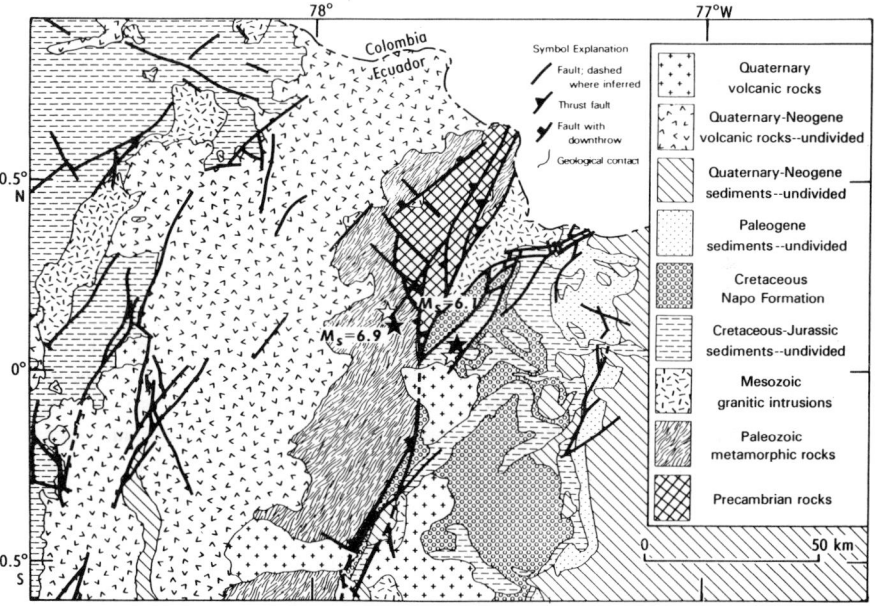

FIGURE 3.2 Map showing the major fault systems in and around the epicentral region. Also shown are the major geological units and the epicenter locations for the main event ($M_s = 6.9$) and of the main foreshock ($M_s = 6.1$) as solid stars and their relocated epicenters as open stars. Their earthquake hypocentral parameters are listed in Tables 3.1 and 3.2.

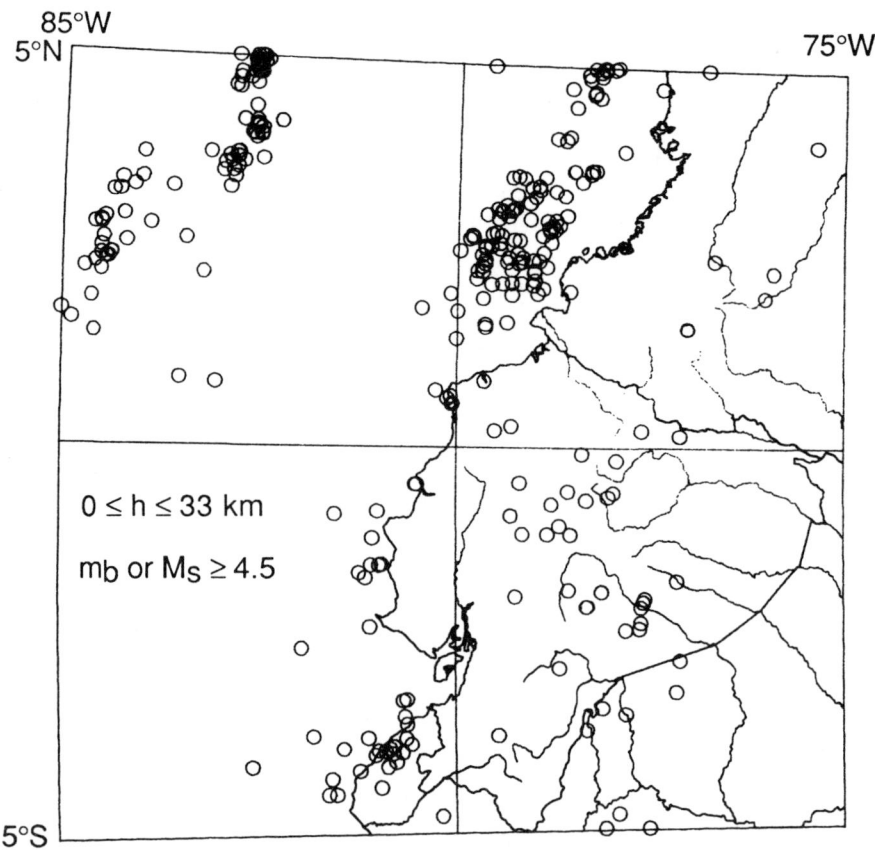

FIGURE 3.3A

FIGURES 3.3 Seismicity for the region of latitude 5°N to 5°S, longitude 75 to 85°W, for the period 1962 to 1987, inclusive. Map projection Oblique Mercator. Epicenters shown have been determined using 10 or more stations for earthquakes with magnitudes (m_b or M_s) equal to or greater than 4.5, and for (A) depth of focus, h, from surface to h equal to or less than 33 km; (B) for h greater than 33 km to h less than or equal to 100 km; and (C) for h greater than 100 km to h less than or equal to 300 km.

Studies of earthquake potential, using conditional probability estimates (Nishenko, 1989), have shown a 66 percent probability for a great earthquake ($M_s > 7.7$) to take place along the subduction zone between latitude 0.5°S and 1.2°N in the recurrence period of 1989-1999. The last large earthquake along this subduction zone occurred on May 14, 1942, with a surface-wave magnitude of 7.9. The conditional probability estimates were evaluated using the historical and instrumental seismicity catalogue of the

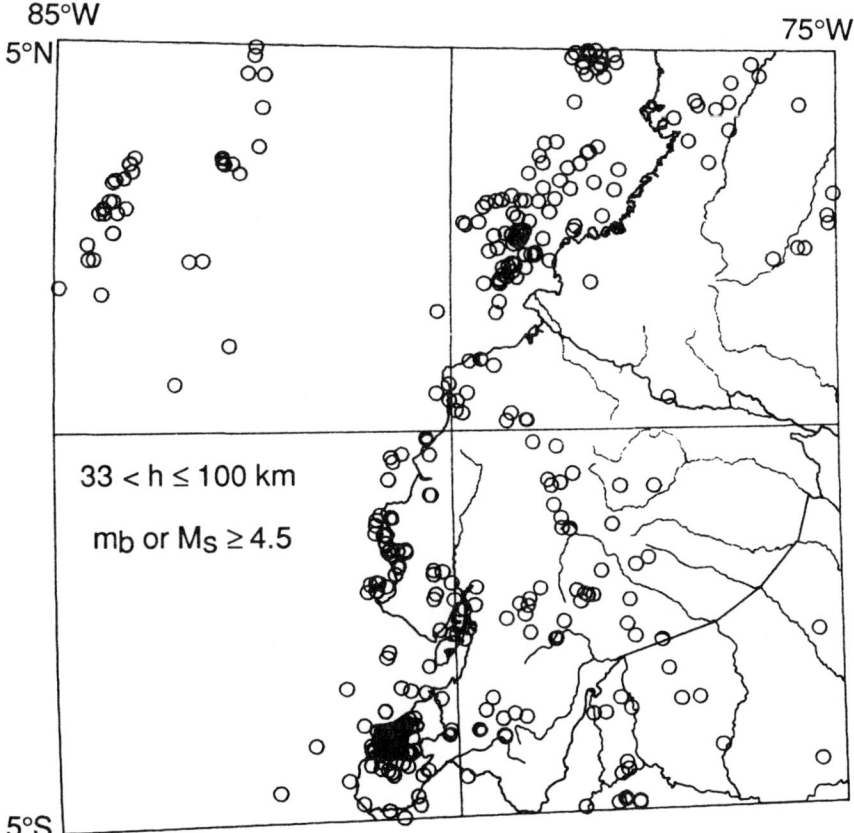

FIGURE 3.3B

region, although the historical record is poor for this region (Heaton and Hartzell, 1986).

The epicenters and magnitudes for the foreshock and the main event of March 5, 1987, determined by the U.S. Geological Survey, National Earthquake Information Center (NEIC), soon after the events, are presented in Table 3.1. These parameters were used to determine the focal mechanism solutions for these earthquakes. The seismic moment (M_o) for the foreshock was 5×10^{26} dyne-cm, and for the main event was 6.4×10^{26} dyne-cm. The above epicenters later were recalculated by the NEIC; the new earthquake parameters are given in Table 3.2.

The locations for the main foreshock, the main event, and the aftershocks were determined using the local network of stations deployed and maintained

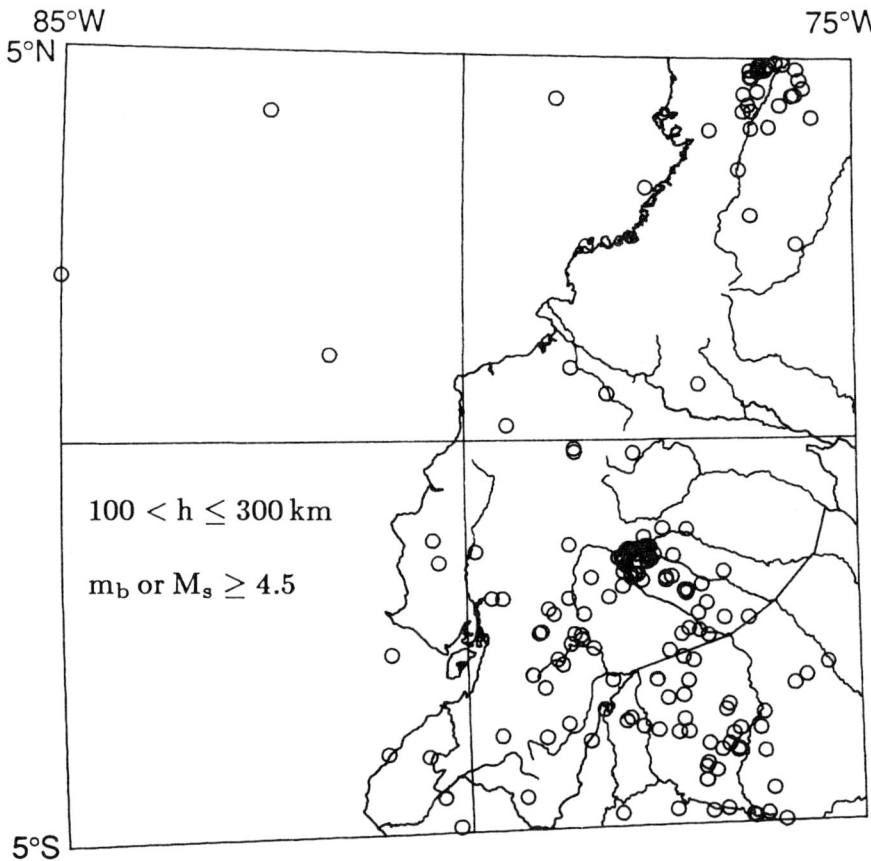

FIGURE 3.3C

by the Instituto Geofísico of the Escuela Politécnica Nacional, Quito; their earthquake parameter determinations are presented in Table 3.3.

Figure 3.2 shows the epicenters of the March 5 earthquakes listed in Table 3.1 as solid stars; their relocated epicenters, as listed in Table 3.2, as

TABLE 3.1 Initial Earthquake Parameters Calculated by U.S. Geological Survey

Date	Time	Lat.	Long.	Depth (km)	No. Obs.	M_s	m_b
5 March 1987	19:54	0.070°N	77.640°W	11	557	6.1	6.1
5 March 1987	22:10	0.120°N	77.800°W	13	594	6.9	6.4

TABLE 3.2 Revised Earthquake Parameters Calculated by U.S. Geological Survey

Date	Time	Lat.	Long.	Depth (km)	No. Obs.	M_s	m_b
5 March 1987	19:54	0.048°N	77.653°W	14	354	6.1	6.1
5 March 1987	22:10	0.151°N	77.821°W	10	344	6.9	6.5

TABLE 3.3 Earthquake Parameters Calculated by the Instituto Geofísico of the Escuela Politécnica Nacional

Date	Time	Lat.	Long.	Depth (km)
5 March 1987	19:54	0.142°S	77.871°W	3
5 March 1987	22:10	0.087°S	77.841°W	12

open stars; and their relation to the regional geology and principal fault systems around the epicentral region.

In Figure 3.4 are shown the number of aftershocks as a function of time following the 19:54 foreshock of the March 5 earthquake. As can be seen, the seismic activity decreased rapidly during the first 38 hr. The number of

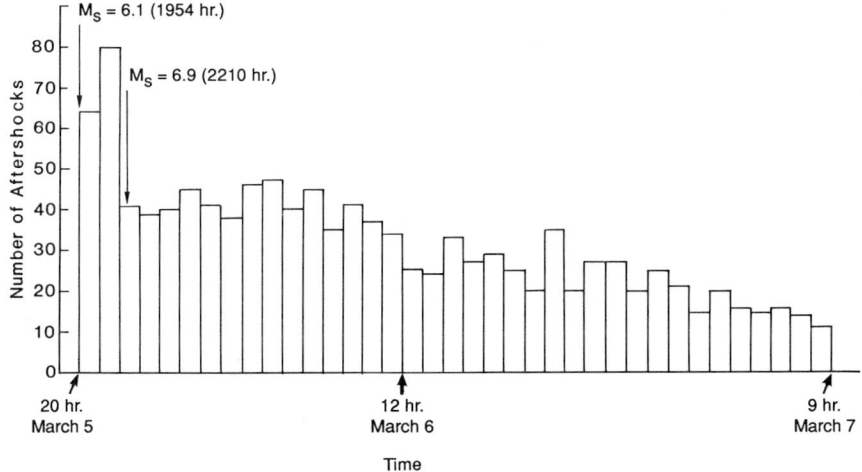

FIGURE 3.4 Number of aftershocks per hr recorded by a field seismological survey from the onset of the March 5, 1987, earthquakes through the first 38 hr. Total number of aftershocks was 1,240.

aftershocks per hr, immediately after the foreshock, was about 64 events, which increased to nearly 80 events before the main earthquake, and then decreased to an average of 40 events per hr during the next 14 hr. The number of events diminished to an average of 30 per hr during the subsequent 13 hr, and further decreased to an average of 15 events per hr during the next 18 hr. The total number of aftershocks in this 38-hr time lapse was 1,240. Relocations were done for 36 aftershocks that occurred between 19:54 and 22:10 on March 5 and for which the S-arrival times could be clearly read. The first group of 19 aftershocks, which occurred between 20:11 and 20:56, fall very tightly on a line with a NW direction, N 49° W. This trend is parallel to the lineament of the Salado River, shown in Figure 3.1. These aftershocks occurred in the region that sustained the highest degree of damage to the environment and to man-made structures. However, the second group of 17 aftershocks, which occurred between 20:57 and 22:03, do not correlate with any of the known faults in the region; however, their epicentral locations are dispersed throughout the meizoseismal area and do not

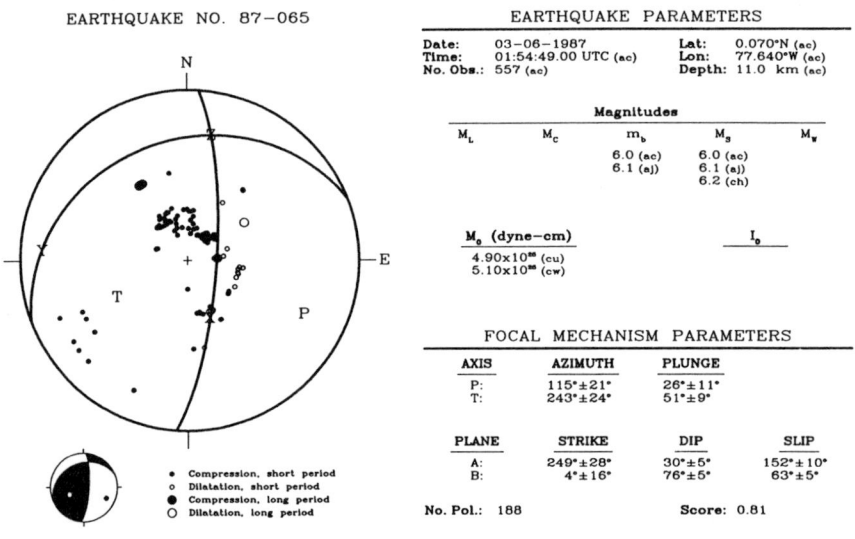

FIGURE 3.5 Focal-mechanism solution for the March 5, 1987, Ecuador earthquakes, with M_s = 6.1, using 188 P-wave observations. The magnitude and seismic moment are listed as obtained by different investigators. The earthquake parameters are the same as those listed in Table 3.1. Under the heading focal-mechanism parameters, the P- and T-axes and the fault planes A and B, Table 3.4, are given, and the uncertainty for each parameter is listed.

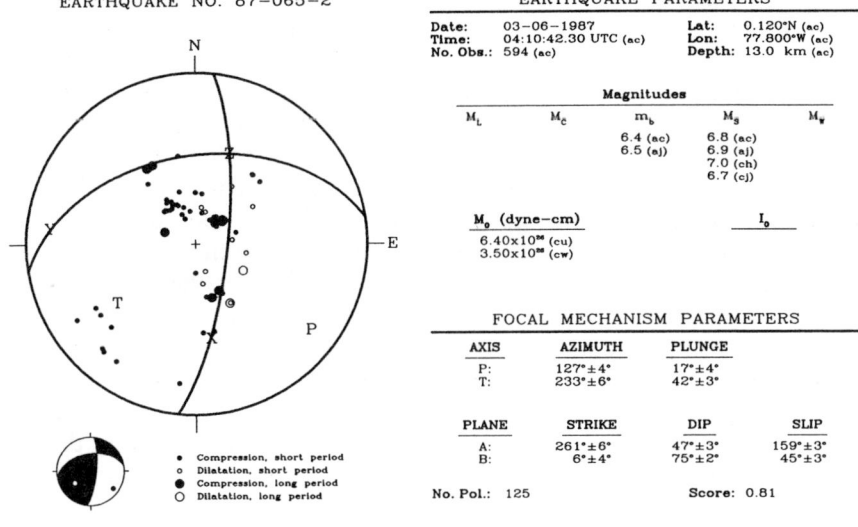

FIGURE 3.6 Focal-mechanism solution for the March 5, 1987, Ecuador earthquakes, with $M_s = 6.1$, using 125 P-wave observations. The magnitude and seismic moment are listed as obtained by different investigators. The earthquake parameters are the same as those listed in Table 3.1. Under focal-mechanism parameters, the P- and T-axes and the fault planes A and B, Table 3.4, are given, and the uncertainty for each parameter is listed.

follow any geologic or tectonic trend. No attempt was made to obtain joint focal-mechanism solutions from the observed aftershocks.

Focal-mechanism solutions obtained from first-motion studies for the foreshock and the main earthquake are shown in Figures 3.5 and 3.6. Table 3.4 lists the fault-plane solutions for these earthquakes. The March 5, 1987, 19:54 and 22:10 earthquakes have very similar focal-mechanism solutions. The uncertainty of each of the focal-mechanism parameters is given in

TABLE 3.4 Earthquake source parameters

	Plane A			Plane B			
EQ. #	Strike	Dip	Slip	Strike	Dip	Slip	No. of Obs.
1	N 69° E	30°W	152°	N 4° E	76°E	63°	188
2	N 81° E	47°W	159°	N 6° E	75°E	45°	125

Figures 3.5 and 3.6. For the main event, the number of polarizations used was 125 and the solution is very stable. The uncertainties in the parameters for the focal-mechanism solution of the foreshock are slightly greater than for the main event. The solutions presented in Figures 3.5 and 3.6 exhibit good correlation with the regional faults shown in Figure 3.2.

The focal-mechanism solutions, shown in Figures 3.5 and 3.6 represent a point-source solution and hence do not represent, in the near field, the finite structure of the rupture process (dislocation). As pointed out in the postearthquake intensity-distribution study by Espinosa et al. (Chapter 4), the town of Olmedo (Figure 4.3), west of the epicenters of the foreshock and main event, sustained higher damage to the infrastructure than did other towns N or S of the epicenters. The isoseismals VI and VII (Figure 4.5 in Espinosa et al., Chapter 4) exhibit a preferential azimuthal distribution to the NW, which could indicate a directivity function due to a dislocation moving in this direction. A similar process, very well defined by the isoseismal distribution, was observed after the Guatemala earthquake of 1976 (Espinosa et al., 1976).

RECOMMENDATIONS

In view of the results we have obtained in the postearthquake field study, we believe there is an urgent need to evaluate the earthquake hazards in Ecuador. We recommend the following activities:

1. Compile a more extensive and detailed historical earthquake catalogue for interplate and intraplate earthquakes in Ecuador.
2. Compile a historical earthquake catalogue for events that are associated with volcanism.
3. Perform paleoseismicity studies in the NW part of Ecuador, especially in the region of Jama (latitude 0.5°S to 1.2°N), in order to ascertain the past occurrence of large and/or great earthquakes in this high-earthquake-potential region.
4. Deploy sensitive seismological instruments in the region (latitude 0.5°S to 1.2°N) in order to ascertain the level of seismicity and to study the joint focal mechanisms.
5. Study the characteristics and ages of marine terraces in order to better understand the past occurrence of great earthquakes in the region.
6. Evaluate the mode of subduction and its geometry in the Wadati-Benioff zone under Ecuador, in order to create a 3-D lithospheric model and to map the maximum horizontal stress distribution in the plate. This task will assist in delineating regions with high earthquake potential.
7. Evaluate the focal-mechanism solutions for all large and great earthquakes in Ecuador.

8. Based upon the above results, outline a working model of the tectonic regime of the region.

REFERENCES

Egred, J. 1988. Terremoto de la Provincia del Napo, Marzo 5, 1987. Instituto Geofísico, Escuela Politécnica Nacional, Quito, Ecuador, 56.

Espinosa, A. F. 1979. Geologic hazards. Pp. 119–144 in Energy Resources of Peru. U.S. Geological Survey Project Report: Peru Investigations (IR) PE-6.

Espinosa, A. F., R. Husid, and A. Quesada. 1976. Intensity distribution and source parameters from field observations of the February 4, 1976, Guatemalan earthquake. Pp. 52–66 in The Guatemalan Earthquake of February 4, 1976, a Preliminary Report. A. F. Espinosa, ed. U.S. Geological Survey Professional Paper 1002.

Hall, M., and C. Wood. 1985. Volcano-tectonic segmentation of the northern Andes. Geology 13:203–207.

Heaton, T. H., and S. H. Hartzell. 1986. Source characteristics of hypothetical subduction earthquakes in the northwestern United States. Bulletin of the Seismological Society of America 76:675–708.

Kanamori, H., and K. C. McNally. 1982. Variable rupture mode of the subduction zone along the Ecuador-Colombia coast. Bulletin of the Seismological Society of America 72:1241–1253.

Lonsdale, P. 1978. Ecuadorian subduction system. American Association of Petroleum Geologists Bulletin 89:981–999.

Lonsdale, P., and K. Klitgard. 1978. Structure and tectonic history of the Eastern Panama Basin. Geological Society of America Bulletin 89:981–999.

Nishenko, S. P. 1989. Circum-Pacific Seismic Potential, 1989-1999. U.S. Geological Survey Open-File Report 89-86:126.

Pennington, W. 1981. Subduction of the Eastern Panama Basin and seismotectonics of northwestern South America. Journal of Geophysical Research 86:10753–10770.

Pilger, R. 1983. Kinematics of the South American subduction zone from global plate reconstructions. American Geophysical Union, Geodynamics Series 11:113–125.

4

Intensity and Damage Distribution

A. F. Espinosa, U.S. Geological Survey, Denver, Colorado
J. Egred, Escuela Politécnica Nacional, Quito, Ecuador
M. García-Lopez, Universidad Nacional de Colombia, Bogotá
E. Crespo, Cornell University, Ithaca, New York

INTRODUCTION

The earthquake of March 5, 1987, was felt at 22:10 local time over an area of at least 93,000 km^2. It originated in the province of Napo in the mountainous region of NE Ecuador; the epicenter was located about 75 km NE of Quito. There were some slight indications of surface faulting; however, the weather and rough terrain precluded a positive field identification of surface-faulting effects. The sense of motion determined from seismic instruments is strike-slip, with a thrust component having an almost N-S strike. A similar source-mechanism solution was obtained from a foreshock and from two aftershocks several months later. The Trans-Ecuadorian pipeline, the nerve center of the Ecuadorian economy, sustained a high degree of damage.

The main event was preceded by a foreshock on March 5, 1987, at 19:54 local time. It had a shallow depth of focus, a body-wave magnitude (m_b) of 6.1, and a surface-wave magnitude (M_s) of 6.1 (Figure 4.1). The main damaging Ecuador earthquake occurred about 20 km W of the foreshock epicenter. It had a body-wave magnitude (m_b) of 6.1 and a surface-wave magnitude (M_s) of 6.9. The maximum Modified Mercalli Intensity (I_o) was IX in the immediate meizoseismal region.

This paper is a preliminary report on the earthquake-damaged area that was studied immediately after the main event took place. The objective was to obtain information, where accessible, in and around the epicentral location, covering a radius of approximately 80 km, in order to delineate the distribution of intensities, damage to adobe-type construction, strong motions, geologic hazards triggered by the earthquake strain-energy release, and related phenomena.

FIGURE 4.1 Epicentral locations of the March 5, 1987, earthquake ($M_s=6.9$) and of the aftershock ($M_s=6.1$) that occurred 3 hr after the main earthquake. Also shown are the departmental divisions (states) of Ecuador.

RELATIONSHIP OF BUILDING DAMAGE TO CONSTRUCTION PRACTICES

In order to assess the intensity distribution from data gathered in the field, it is necessary to identify the types of building construction in the areas examined.

To the E of Quito, the region best known as the *región oriental* (eastern region), the predominant type of building construction is wood-frame. There is also a mixture of construction practices in this region; for example, wood-frame and brick combination, also wood-frame and cinder-block mixture, and to a lesser extent, the replacement of brick by adobe. In some of the important towns, as well as in the petroleum exploration camps, the structures are of reinforced concrete, usually no more than four stories high. Most roofing in the larger towns is of corrugated zinc-plated steel; the use of terra-cotta tiles is very limited. Nearly all structures in rural areas and small hamlets are single-story stone buildings. For this eastern region of Ecuador, a preliminary count of structures that sustained damage in the towns of Baeza, Lumbaquí, El Chaco, Reventador, Lago Agrio, and Díaz de Pineda (Figures 4.2 and 4.3) totaled 1,976; 818 were damaged beyond repair (as of May 13, 1987).

To the N, in Pichincha, Imbabura, and Carchi (Figure 4.1), the predominant type of house construction is known as *tapial*, in which the dwellings have no reinforcement on their corners, nor do they commonly possess interior walls or reinforced columns. The most common construction materials for walls are *adobe* and *bahareque* (adobe is a mixture of mud and straw; bahareque is a mixture of mud and long sticks of cane sugar or some

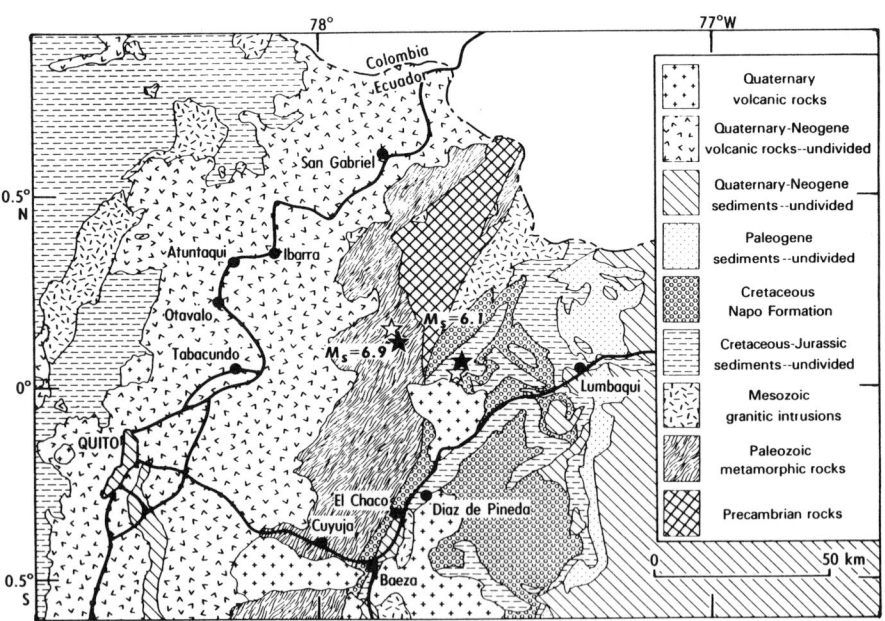

FIGURE 4.2 Map showing the major geological units for the epicentral region. Also shown are the epicenters of the main event ($M_s=6.9$) and of the main aftershock ($M_s=6.1$), and the main access highways to the E and N of Quito.

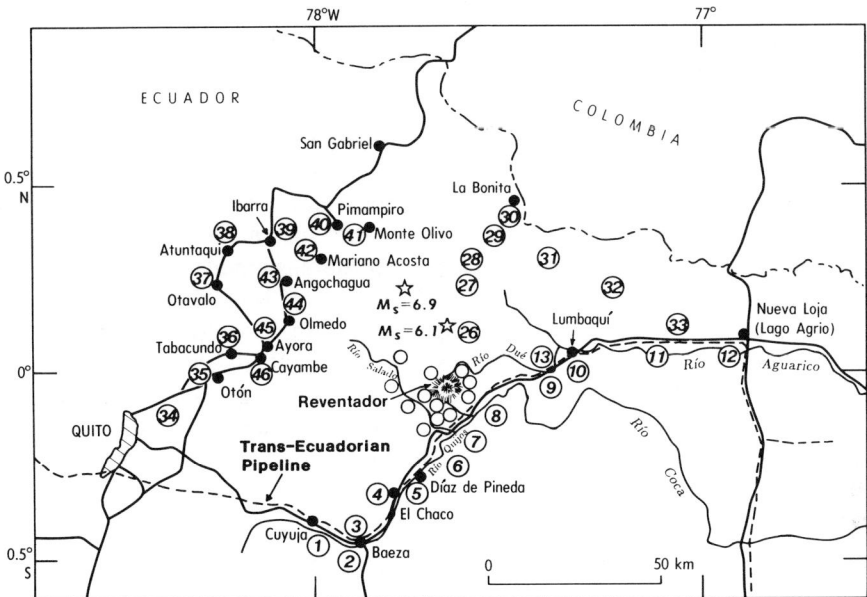

FIGURE 4.3 Intensity sampling distribution obtained in this postearthquake study around Reventador Volcano and along the main access roads to the N and E of Quito. Also shown are those areas in which an intensity evaluation was performed from helicopter reconnaissance flights. The circled numbers represent the areas where damage to dwellings and to the environment was documented using a video camcorder and 35-mm photographs.

similar material). Fewer buildings are constructed of brick and/or cinder blocks. In this part of Ecuador, known as *región interandina* (northern region), it is customary to cover the roof with terra-cotta tiles. The walls are quite thick; some are about 1 m wide, and in some of the old historic churches they are even thicker.

The larger towns and cities of Ecuador employ a great variety of types of construction, from adobe to reinforced-concrete buildings several stories high. The ages of the structures also vary greatly.

As of May 13, 1987, the number of dwellings that sustained damage in the region interandina totaled 12,083; 2,308 structures were damaged beyond repair. For example, in the town of Olmedo (Figure 4.3), due W of the March 5 epicenters, 1,631 dwellings sustained damage beyond repair.

The towns of Cangahua and Ayora (Figure 4.3), just N and S of Olmedo, sustained damage to 1,300 and 323 dwellings, respectively; 570 and 302 of these dwellings were damaged beyond repair. In the towns of Tabacundo and Cayambe (Figure 4.3), which are SW of the epicentral locations, much less damage occurred. To the NW, much higher per capita damage occurred in Ibarra, where about 814 dwellings were damaged beyond repair. Farther N, at San Gabriel (Figure 4.3), damage decreased to 220 structures that were damaged beyond repair.

The damage distribution from this earthquake yields a rather uniform picture if, and only if, we study the same kinds of structures. Similar findings were field-documented by Espinosa and others (1976b) after the February 4, 1976, Guatemala earthquake. In the Guatemala event, the mapped infrastructure damage distribution reflected the sustained damage to adobe construction, or the equivalent of a rating in the MMI scale of VII. In Guatemala, the distribution of damage to buildings of adobe construction appeared to be directly related to the intensity (MMI) distribution (Espinosa et al., 1976b). In Ecuador, in the area of heavy landsliding, many adobe and wood-frame houses sustained no damage. Numerous scattered dwellings near landslides along the main highway to the east of the affected region were not damaged.

INTENSITY DISTRIBUTION

The authors visited villages in areas of high, intermediate, and low damage, and, by using questionnaires and a video camcorder, gathered data used to assess the MMI ratings throughout the affected region (Figures 4.3 and 4.4). In these figures, the circled numbers identify those areas where damage to man-made structures and to the environment were photographed and certain areas where videotapes were taken by the authors documenting the aftereffects of these damaging earthquakes.

The areas of maximum MMI are concentrated in the meizoseismal area, attaining an I_o=IX (Egred, 1988; García-Lopez, 1988). In this area, however, much of the damage can be classed as intensities VII and VIII. The problem encountered in the process of evaluating the intensity is that of "inconsistencies" in the MMI and in the Medvedev-Sponhauser-Karnik (MSK) intensity scales. Large landslides, such as were abundant in the high mountainous region near and around Reventador Volcano, as well as in other unstable regions in the high Andes (Nieto et al., Chapter 5, this report), suggest an intensity greater than IX. Another factor that indicates high intensities (>IX) is surface faulting, examples of which were observed in the Cascabel location (near circled number 25, Figure 4.4), near the epicentral region. Still another factor that yields higher intensities (X), as given in the intensity scale, is "landslides considerable from river banks and steep slopes due in some cases to water-saturated soils, shifted sand and mud, water splashed over banks." Also, "bridges destroyed" implies an intensity rating of XI, and although a bridge was indeed destroyed (but by flooding), wooden structures nearby sustained no damage at all. Similar problems in

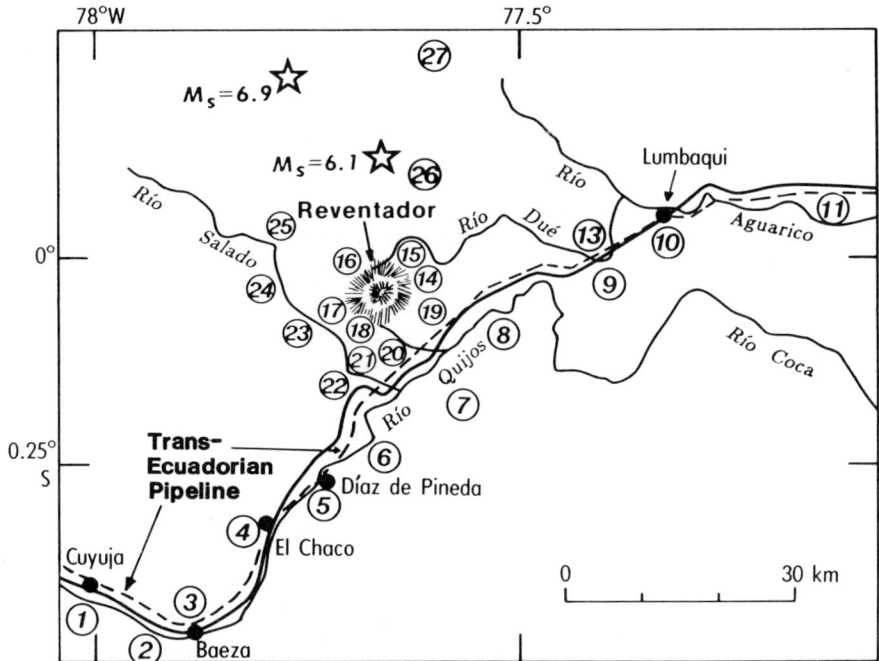

FIGURE 4.4 Intensity sampling distribution detail around Reventador Volcano and the region around the Salado River, the Quijos River, and part of the Aguarico River. Identification of circled numbers is explained in the caption of Figure 4.3.

the inconsistencies inherent in the MMI have been documented for several postearthquake studies in Caracas, Venezuela; Lima, Peru; Guatemala; and El Asnam, Algeria (Espinosa, 1969, 1976, 1981; Espinosa and Algermissen, 1972, 1973; Espinosa et al., 1976a, 1976b).

The data obtained in questionnaires and field notes gathered after the March 5 earthquakes have been rated by using the abridged version of the MMI scale (Richter, 1958) with the following exceptions: landslides are not considered in our report as indicators of intensity X; "ground cracked conspicuously" and "underground pipes broken" are not considered as indicators of intensity IX; "shifted sand and mud, as well as water splashed over banks" are not considered as indicators of intensity X; "broad fissures in ground and bridges destroyed" are not considered as indicators of intensity XI. Espinosa and others (1976b) have proposed that these exceptions and other ground-failure effects be considered second-order effects that do not represent the intensity of an earthquake based on purely vibrational effects. A similar proposal is being implemented by the Soviet Intensity

Scale Committee in the revision of the MSK intensity scale (L. Shebalin, 1986, personal communication; D. Mayer Rosa, 1988, personal communication). Other secondary effects, such as liquefaction, ground compaction, subsidence, surface breakage, massive landslides, rock slides, debris flows, and "damming" of rivers, were observed in the field in the vicinity of Reventador Volcano and the tributary rivers in the region. According to the MSK intensity scale, the ratings, if followed verbatim, would suggest much higher intensities than those observed and assigned in the field and based on the vibrational effects produced by the earthquake.

Figure 4.4 shows the location of Reventador Volcano and the areas mentioned above that sustained high damage to the environment. The number of landslides in this area around and on the volcano was very high, and several large landslides that most likely were caused by water-saturated soils occurred E of the epicentral region; however, very few occurred to the N.

The isoseismal in Figure 4.5 shows a preliminary distribution of the main earthquake in Ecuador in March 1987. The isoseismals for intensities VIII, VII, and VI follow the general trend of a system of faults trending N-S. The intensities attenuate over a shorter distance more rapidly to the SW, and more slowly to the S and to the NNW. The isoseismal map was plotted on a 1:1,000,000 scale geologic map of Ecuador (Baldock, 1982a,b), and no simple correlation was found between gross surficial geology and intensity distribution.

The focal-mechanism solution obtained by Espinosa et al. (Figures 3.5 and 3.6) for the main event has a strike for fault planes A and B of 261 and 6°, respectively. The isoseismals VII and VI exhibit a lobe to the NW that could indicate an effect due to a moving source in the near field. There is not enough intensity-data resolution in that region, because of the inaccessibility at the time of the earthquake, to ascertain or to infer the direction of the maximum released energy at those azimuths. The shaking intensity for MMI=VII has a certain directivity component suggesting a unilateral rupture from E to W. As pointed out earlier, the town of Olmedo (Figure 4.3) sustained higher damage to its infrastructure than municipalities N or S of Olmedo.

Other factors that may enter into the intensity distribution pattern shown in Figure 4.5 are seismic amplification effects, topographic seismic wave amplification, influence of surficial soil conditions, and depth of the water table. Moreover, construction practices in this part of the country are highly mixed, and in the mountainous area the population settlements are small and scattered.

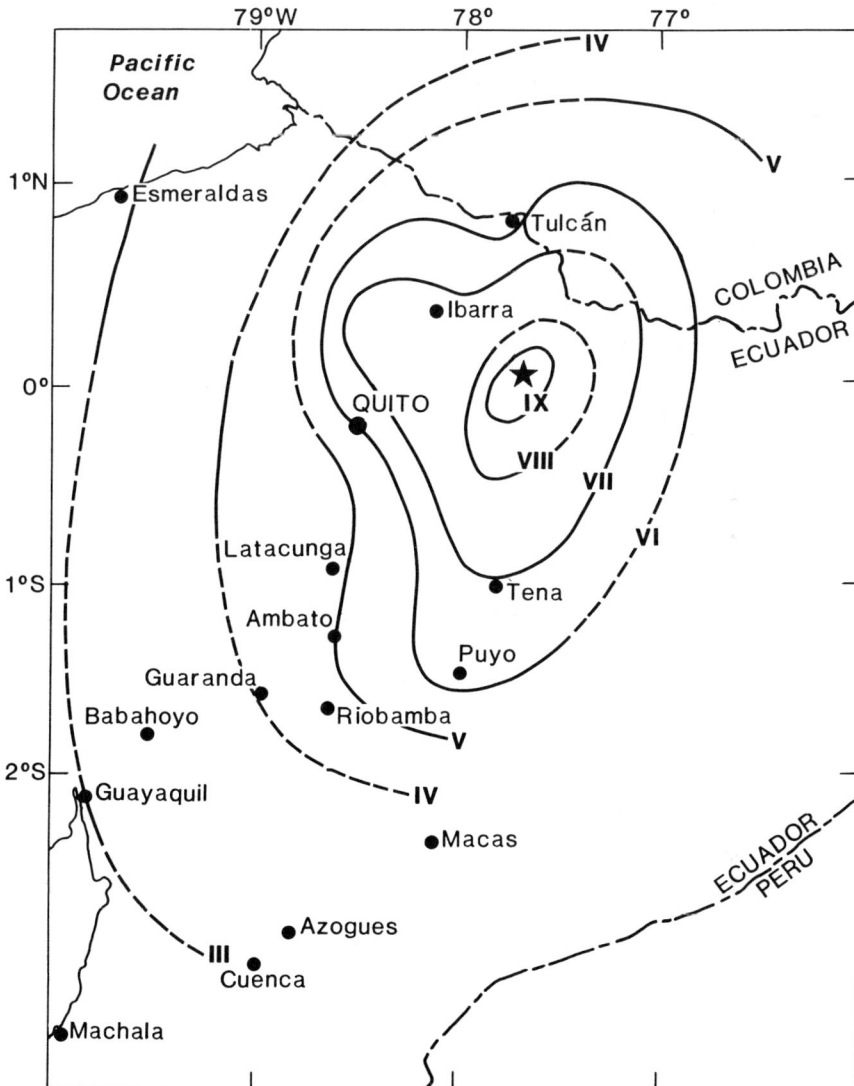

FIGURE 4.5 Intensity distribution in Ecuador from the main earthquake of March 1987. The solid star indicates the epicenter. Dashed lines indicate approximate isoseismic extension or continuation. Also shown are the locations of important cities and towns in Ecuador.

REFERENCES

Baldock, J. N. 1982a. Geology of Ecuador—Explanatory Bulletin of the National Geological Map of the Republic of Ecuador. Ministerio de Recursos Naturales y Energéticos, Dirección General de Geología y Minas, Quito, 54 plus references.

Baldock, J. N. 1982b. Geological Map of Ecuador. Ministerio de Recursos Naturales y Energéticos, Dirección General de Geología y Minas, Quito, scale 1:1,000,000.

Egred, J. 1988. Terremoto de la Provincia del Napo, Marzo 5, 1987. Instituto Geofísico, Escuela Politécnica Nacional, Quito, 56.

Espinosa, A. F. 1969. Ground amplification of short-period seismic waves at two sites near Bakersfield, California. Earthquake Notes 40:3–20.

Espinosa, A. F. 1976. The Guatemalan Earthquake of February 4, 1976. U.S. Geological Survey Professional Paper 1002, 90.

Espinosa, A. F. 1981. The Algerian earthquake of October 10, 1980—A preliminary report. Earthquake Information Bulletin 13:23–33.

Espinosa, A. F., and S. T. Algermissen. 1972. A Study of Soil Amplification Factors in Earthquake Damage Areas, Caracas, Venezuela. Environmental Research Laboratories TR-280-ESL-31, 201.

Espinosa, A. F., and S. T. Algermissen. 1973. Ground amplification studies in the Caracas Valley and the northern coastal area of Venezuela. Proceedings of the Fifth World Conference on Earthquake Engineering, Rome 2:106.

Espinosa, A. F., R. Husid, S. T. Algermissen, and J. de las Casas. 1976a. The Lima earthquake of October 3, 1974, intensity distribution. Bulletin of the Seismological Society of America 67: 1429–1440.

Espinosa, A. F., R. Husid, and A. Quesada. 1976b. Intensity distribution and source parameters from field observations of the February 4, 1976, Guatemalan earthquake, in The Guatemalan Earthquake of February 4, 1976, a Preliminary Report, A. F. Espinosa, ed. U.S. Geological Survey Professional Paper 1002, 52–66.

García-Lopez, M. 1988. Evaluación de los problemas de inestabilidad del terreno causados por los sísmos del 5 de Marzo de 1987. Bogotá: Universidad Nacional de Colombia, 41.

Richter, C.F. 1958. Elementary Seismology. San Francisco: Freeman, 768.

5

Mass Wasting and Flooding

A. S. Nieto, Department of Geology, University of Illinois, Urbana
R. L. Schuster, U.S. Geological Survey, Denver, Colorado
G. Plaza-Nieto, Escuela Politécnica Nacional, Quito, Ecuador

INTRODUCTION

This chapter deals with slumps and slides, debris avalanches, debris flows, and flooding, the processes that account for the largest amount of destruction and number of deaths induced by the March 5, 1987, earthquakes.

LANDSLIDES

Landslide processes are described first, then interpreted in light of the field observations. The descriptions are grouped into three field traverse areas (from W to E) that are progressively closer to the epicenters of the earthquakes, located about 25 km NNW of Reventador Volcano: Quito-Baeza, Baeza-Salado, and Reventador Volcano (Figure 5.1). Field observations for the first two areas were made mostly by land along the Trans-Ecuadorian highway from Quito to the mouth of the Salado River; the observations in the Reventador area were made by car, helicopter, and airplane.

Quito-Baeza

This traverse is divided into three approximately equal parts, each about 25 km long. Midpoints of each of these sections are 90 km, 80 km, and 75 km from the epicenters. The first section is the eastern half of the Inter-Andean Valley, where Quito is located. This depression is filled mostly with pyroclastic materials—primarily "cangahua"—and glacial till (see Chapter 2). Elevations along this section range from 3,000 to 3,500 m, and the climate is temperate. Slopes are relatively gentle, generally not exceeding 20 to 25°. A few slope failures caused by the March 5 earthquakes were

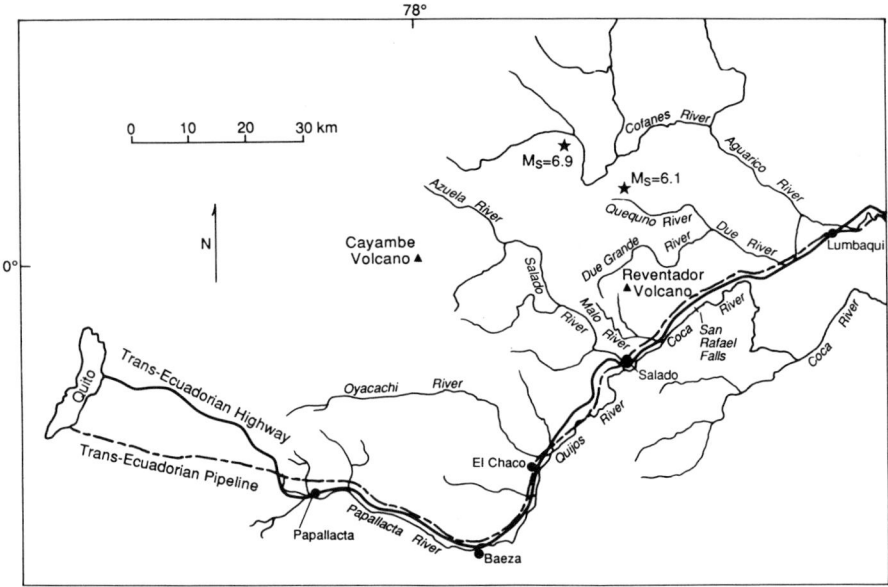

FIGURE 5.1 Map of area E of Quito showing locations of epicenters of the March 5, 1987, earthquakes, the Trans-Ecuadorian pipeline and highway, Reventador and Cayambe volcanoes, and rivers and towns noted in this chapter.

observed in cangahua along very steep to subvertical highway cuts and stream banks. Steep slabs or overhangs failed as falls, topples, and slips. (Mechanical properties and failure mechanisms in the cangahua have been described by Crespo and Stewart [1987]). The types of failures—falls and topples—occurring the farthest from the epicenter confirm observations by Keefer (1984). We also observed fresh slumps and soil flows in sandy morainal deposits at the headwaters of the Papallacta River (Figure 5.1), but ascertaining if these were caused by the March 5 earthquakes is difficult. Further, we observed a couple of examples of old landslide morphology that showed no indication of reactivation.

The next 25 km crosses lava flows that create the divide of the Cordillera Real (Figures 2.1 and 2.2). The lava flows are Quaternary rhyolites, andesites, and basalts. The elevation is about 4,000 m, and the climate is temperate to cold. Relief is high, and soils are nonexistent or colluvial. Slopes are nearly vertical in the lava flows, and moderate—25 to 35°—in talus deposits at the foot of cliffs. Some slopes have columnar joints or stress-relief joints. We observed a few topples, some of which may have been related to the earthquakes, and also a few talus slopes with evidence of some recent movement.

MASS WASTING AND FLOODING

The last 25 km or so to Baeza comprise the descent to the sub-Andean zone (western edge of the Oriente, Figures 2.1). In this stretch, the highway follows the Papallacta River. The rocks are Cambrian and Precambrian gneisses and schists. Elevations range from 3,500 m at the western end to 1,500 m near Baeza. Relief is very pronounced; some slopes are 45° or steeper. Residual soils are thin or absent for most of the section. Only in the lower reaches of the Papallacta River, a few kilometers E of Baeza, did we observe a few debris flows on the highest parts of the Papallacta valley walls. These debris flows were shallow and had the same characteristics as those described in detail in later sections.

Baeza-Salado

This traverse extends from the town of Baeza down the Quijos River, where it joins the Salado River to form the Coca River (Figure 5.1). In general, the denudation around Baeza was only moderate. Some shallow debris flows in residual soils could be seen on the crests of slopes around the town. In addition, a small number of rock slides and rock glides, as well as a few slips and slumps in old and recent alluvial terraces (Figure 5.2), were observed. Landslide intensity increased progressively from El

FIGURE 5.2 Thin earth slide (center of photo) on steep face of alluvial terrace on western edge of the town of Baeza.

Chaco to Salado. The pattern of landsliding included initiation points located high on steep slopes covered with shallow residual soil and weathered rock. This pattern, which became apparent in areas closer to the epicentral zone, is discussed in detail below. The most significant observation along this traverse is that very few of the debris flows reached the Quijos River channel. Thus, landslide material contributed little to the sediment load of the Quijos. Some detailed observations along this last portion of our traverse follow:

Young alluvial terraces, 15 to 20 m high near Baeza, have been created by stream degradation (Papallacta River and tributaries). Intense chemical weathering has modified these alluvial deposits, giving them certain characteristics of residual soils. This veneer of residual/alluvial materials commonly was involved in slip/debris-flow failures of the steep terrace slopes. The relatively unweathered alluvial material (probably cemented) underneath this veneer remained unaffected.

From Baeza to about 20 km to the NE, the Quijos River follows the contact between Paleozoic and Mesozoic rocks (Baldock, 1982), probably coincident with a major thrust; the Trans-Ecuadorian highway and pipeline lie on the W (left) side of the Quijos valley (Figure 5.1). For the next 20 km to the confluence of the Quijos and Salado rivers, the highway and pipeline begin to climb over Cretaceous sedimentary and metasedimentary rocks. These lifelines follow high elevations and drainage divides, which, in turn, coincide with hogbacks formed by generally massive quartzose sandstones that occur within the Mesozoic sequence. Note that the residual soil veneer on the steep slope of the hogback is very thin or nonexistent, as is generally the case in this area. Consequently, the earthquake-induced landsliding was also very shallow. Slope failure began as slips at the highest points of the steep slopes; the slip surfaces coincide with the top of weathered rock and the slips moved rapidly downslope, becoming debris flows (Figure 5.3). This type of landsliding endangered the highway and pipeline, but the slips were generally very thin (less than 1 m) and seldom reached the channel of the Quijos (Figure 5.4).

From about 30 km NE of Baeza to the confluence of the Coca and Salado rivers, the highway follows the steep slope of a massive hogback; for most of this distance the highway is closely paralleled by the pipeline. The highway has been located approximately at the contact between the steep upper slope, covered with a very thin veneer of residual soil, and the lower, gentler colluvial slope. Many very shallow rock falls, debris avalanches, and debris flows (Figure 5.5) moved down from above the highway and blocked it. Although these landslides blocked the highway for several kilometers, local volumes of weathered rock and soil were small, and the landslide materials were fairly easily removed by maintenance crews.

FIGURE 5.3 Landslides (largest examples are indicated by arrows) in residual soils at the top of the N valley wall of the Quijos River between El Chaco and the confluence of the Quijos and Salado Rivers. Because slide-resistant sedimentary bedrock (light-colored outcrop at the lower left) is close to the surface, these slides/avalanches are shallow and generally limited to existing gullies. Although the slides endangered the Trans-Ecuadorian pipeline and highway (angling across top of photo above the landslides), they caused little actual damage because they were so shallow.

Reventador Volcano

Geological Background

This area includes the greatest intensity of landsliding triggered by the March 5 earthquakes. The Reventador Volcano zone, as defined here, is located in the sub-Andean region; it centers on Reventador Volcano and is bounded by the valleys of the Salado River on the W, the Coca River on the SE, the Quequno River on the N, and the Dué River on the NE (Figure 5.1). The zone includes the greatest intensity of landsliding triggered by the March 5 earthquakes. The earthquake epicenters lie a few kilometers to the N and W of the Reventador area; they are not immediately adjacent to the areas of most intense landslide activity.

The zone presents a large amount of relief both as a unit and locally; elevations range from about 1,550 m on the floor of the valley of the Coca

FIGURE 5.4 Landslide activity on the NW (left) valley wall of the Quijos River just upstream (SW) from its confluence with the Salado River. The earthquake-triggered slope failures began as thin slips on the steep (45 to 50°) headwalls of the slopes, and were transformed into debris avalanches and debris flows that cascaded down the gullies onto the low terrace that forms the left bank of the river.

River to 3,560 m at the top of Reventador Volcano. The ridges on the W side of the Salado River rise to 3,600 m, and the ridges on the E reach 3,200 m; the midreaches of the Salado River valley are at an elevation of about 1,600 m. Drainage patterns are radial on the volcano slopes and dendritic or rectangular elsewhere. The rectangular pattern reflects the fracture systems that affect this region.

Reventador Volcano and related volcanic deposits constitute the most notable morphologic feature of the zone (Hall, 1977). This feature is made up of a portion of a large cone—the remains of the collapse of two ancient Reventador stratovolcanoes called Paleo Reventador I and II—that contains a smaller cone, the modern Reventador Volcano. The ancient cone resembles an amphitheater, which opens to the E toward the Coca River.

The overall slopes of the uppermost portions of the ancient cone have angles of about 30 to 35°; the slopes decrease progressively to 10° or less in the lowermost portions. The W and SW slopes of the ancient cone have two important and unique morphologic characteristics. The portion of the volcano between the headwaters of Morales Creek (a tributary of the Malo

River) and the Dué Grande River (Figure 5.1) has a dense parallel drainage pattern; long, closely spaced, parallel gullies have deeply dissected the underlying pyroclastic rocks. These gullies are from 50 to 100 m deep and have gradients that commonly are greater than 35 to 40°. The canyons of Morales Creek, the Dué Grande River, and two unnamed tributaries to the N of the Dué Grande have valley walls that reach heights of more than 200 m and have slope angles greater than 60°. The headwaters of these canyons are cirque-like and present large piping (internal-erosion) cavities in the pyroclastic beds. The valley walls of the Coca River have overall slopes generally between 30 and 40°. The top of the right valley wall, which has been eroded into quartzose sandstone of the Hollin Formation (geologic map, Figure 5.6), is almost vertical in several places. Downstream from San Rafael Falls (Figures 5.1 and 5.7), the Coca River becomes entrenched, and the walls are very steep or vertical. Gullies descending the main valley slopes of the Coca River have gradients of 45 to 60°. The valley walls of the Malo River commonly are steeper than 45°. The main valley walls of the

FIGURE 5.5 Landslides along the Trans-Ecuadorian highway 3 km W of its crossing of the Salado River. These rock falls, slides, and avalanches, which were triggered by the March 5, 1987, earthquakes, had blocked the highway almost continuously in the stretch shown in this photograph. The highway was passable at the time of this April 25, 1987, photograph because of the efforts of Corps of Engineers (Ecuadorian Army) highway maintenance crews.

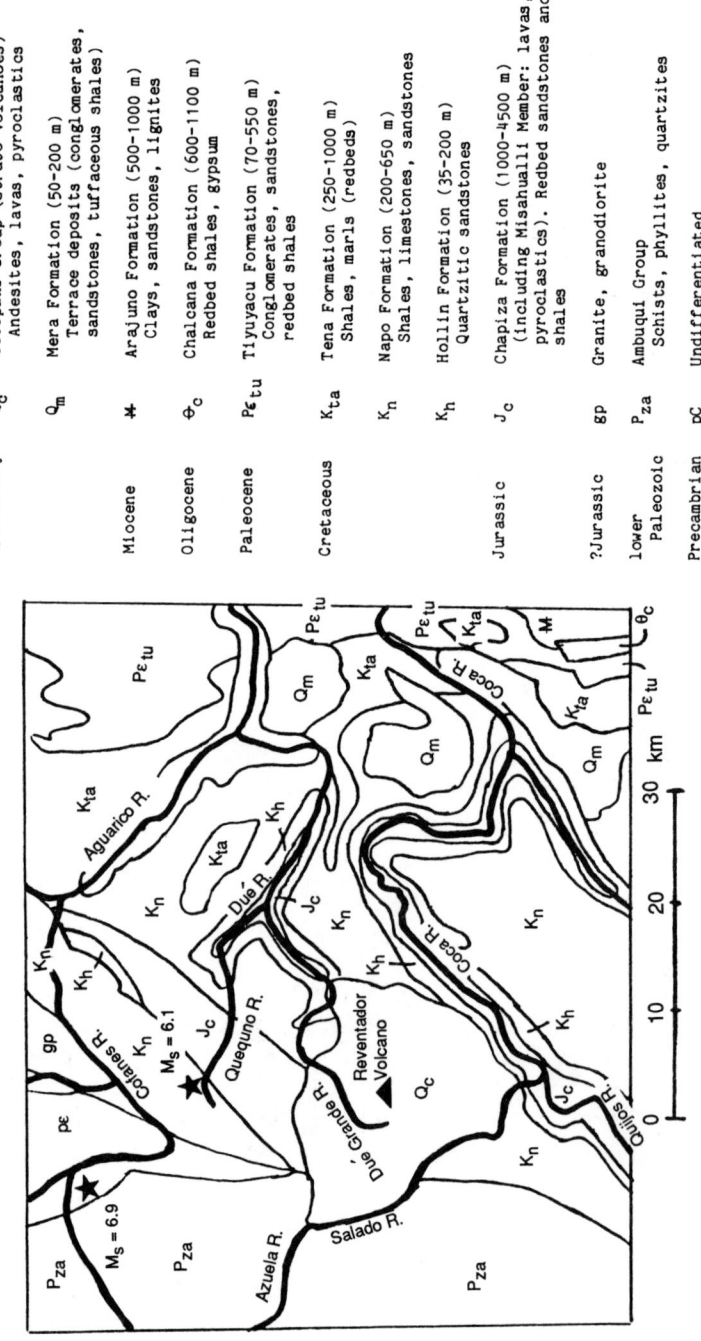

FIGURE 5.6 Geologic map of the Reventador Volcano area (modified from Baldock, 1982).

FIGURE 5.7 April 1987 aerial view of San Rafael Falls on the Coca River downstream from Reventador Volcano. Note the vegetation trim line (arrows), which indicates the maximum height of the debris flow/flood, about 20 m above the current level of the Coca River. There is a strong probability that constriction of the river at the falls caused short-lived "hydraulic" damming of the river.

Salado River have overall slopes that usually are between 25 and 35° in their middle reaches. Tributaries of the Salado have variable overall slope angles, about 40° in their midcourses but lower values in their lower courses and uppermost reaches. The valley/gully walls of first- and second-order tributaries of the Salado have slope angles that range from 40 to 60°, or even greater.

The zone has distinct microclimates that vary from cold and dry (annual precipitation of less than 1,500 mm in the highest portions), to temperate in the tributaries of the major rivers, to subtropical (annual precipitation of more than 4,000 mm along the valleys of the lower parts of the Salado, Coca, and Dué rivers). Consequently, the vegetation is a function of elevation. Oaks and some evergreens grow at the highest elevations, shrubs and low trees in the middle elevations, and rain-forest vegetation below 2,500 m.

Rainfall occurs throughout the year, but increases in intensity from March to July. Significantly, anomalously high precipitation occurred in the area in January and particularly in February 1987. On February 3 and 20, the

INECEL gaging station just upstream from San Rafael Falls (Figures 5.1 and 5.7) on the Coca River registered flow rates of 2,600 and 3,400 m^3/sec, respectively. These discharges have estimated return periods of 5 and 20 years, respectively. A few days before the earthquake, the flow rates diminished. Thus, the gaging station measured 450 m^3/sec at 1700 on March 5; the estimated discharge at 2300 was 600 m^3/sec. These values, however, are more than twice the average flow of the Coca River.

Most of the Reventador zone is underlain by subhorizontal and gently dipping sedimentary rocks of Jurassic/Cretaceous and Cretaceous ages (Figure 5.6). The formations mapped in the Reventador zone are from the oldest to the youngest: pyroclastic sedimentary rocks (Misahualli Member of the Chapuzi Formation), quartzitic sandstones (Hollin Formation), shales and limestones (Napo Formation), and red clay shales (Tena Formation). The western boundary of the Reventador zone is composed of Paleozoic gneisses and schists in contact with the sedimentary rocks to the E by major overthrusts. The sedimentary rocks adjacent to the overthrusts have undergone low-grade metamorphism along a belt a few kilometers wide that extends close to the western edge of the Reventador volcanic complex. In the vicinity of Reventador Volcano, the rocks are lavas and pyroclastics, ranging in age from late Pleistocene to Holocene.

Rectangular drainage in some areas and lineaments observable in airphotos provide evidence of strong fracturing. This fracturing affects even the volcanic materials of Holocene age. The most important fracture systems trend N-S and NE-SW. The latter may be responsible for SW-NE course of the Quijos/Coca River.

Away from the flood plains and alluvial terraces of the main rivers and high-order tributaries, the soils are either colluvial or residual. The colluvial soils occur at the foot of the slopes and were generally not involved in the March 5 landsliding. The residual soils range from saprolites to laterites. Because the relief is significant and the area is well drained, the residual mantle is not very thick—from a few centimeters to a few meters at most—and grades rapidly into weathered fractured rock. These soils have very high void ratios and natural water contents; the values of the latter are practically equal to the liquid limits. Under these conditions, these soils have sensitive structures.

General Characteristics of Landslides in the Reventador Zone

More than 90 percent of the observed landslides began as shallow slips or slides of residual soils and highly weathered rock on the uppermost parts of the slopes of the main valleys or on the slopes of the lower-order tributaries (Figure 5.8). Average thicknesses of these slips were from 1.5 to 2.0 m, with a thickness range of a few decimeters to 5 m. The failing masses

MASS WASTING AND FLOODING 61

FIGURE 5.8 Ridge NW of Reventador Volcano from which jungle vegetation has been stripped by landslides triggered by the earthquakes. The slope failures started as thin slips on the steep (about 50°) upper slopes. As they moved downward, the saturated, vegetation-charged masses were transformed into debris avalanches and then to debris flows that discharged into the Dué Grande River (out of the photo at the bottom). (Photograph by Ken Nyman, Cornell University).

either were transformed into debris avalanches and then into debris flows or, in some cases, were reworked almost immediately into debris flows with high fluidity. Whether they began on a main valley slope or the valley slope of a lower-order stream, flows moved to the channels of the lower-order tributaries and entrained the colluvial/alluvial fills along the channels, eroding the material to the bedrock surface. The debris flows then proceeded to the higher-order tributaries or to the main river valleys. Smaller debris flows stopped lower on the slopes of the main river valleys or in the bottoms of gullies. Debris flows starting higher above the valley floor were also more apt to reach the floodplain than those from lower heights. A large number of the landslide scars displayed unweathered bedrock, attesting to the shallowness of the residual soil mantle.

Overall preearthquake slope angles and gully slope angles on topographic maps at locations of the earthquake-induced slope failures were measured; the locations were determined from postearthquake vertical and oblique airphotos. These measurements indicated that on the left valley wall of the

Coca River, between the Salado pumping station and the Malo River, failures occurred mainly on slopes steeper than 35 to 40°. The same threshold values were obtained for the valley walls of the Salado River and of the Coca River at the confluence of the Salado and Quijos rivers (Figures 5.9 A,B,C). However, on the upper slopes of the ancient cone of Reventador Volcano at elevations above 3,500 m, we observed failure on slopes between 30 and 35°, involving what appear to be very recent ashes. Other types of landslides (rock avalanches, rock slides along high road cuts, slumps and slides along alluvial terraces, and topples of stress-relief slabs) occurred throughout the entire area, but only sporadically.

Areal Distribution of Landslides and Denudation Intensity

Figure 5.10 shows the approximate limits of areas with different degrees of denudation (expressed as estimated percentage of failed slope area per total area) for the slopes of Reventador Volcano and vicinity. Note the large variation in denudation intensity around the volcano. The degree and characteristics of denudation due to landsliding in nearby valleys are as follows:

FIGURE 5.9A 1978 view showing the jungle-covered right (SE) valley wall of the Coca River at the confluence (photograph by S. D. Schwarz).

FIGURES 5.9A, B, C Views down the Salado River to the confluence of the Salado and Quijos Rivers to form the Coca River.

FIGURE 5.9B April 1987 view of the same valley wall showing widespread denudation due to March 5 landsliding.

FIGURE 5.9C June 1990 view of the same valley wall showing partial recovery of vegetation on 1987 landslide scars. Note reconstructed Trans-Ecuadorian highway bridge across the Salado River.

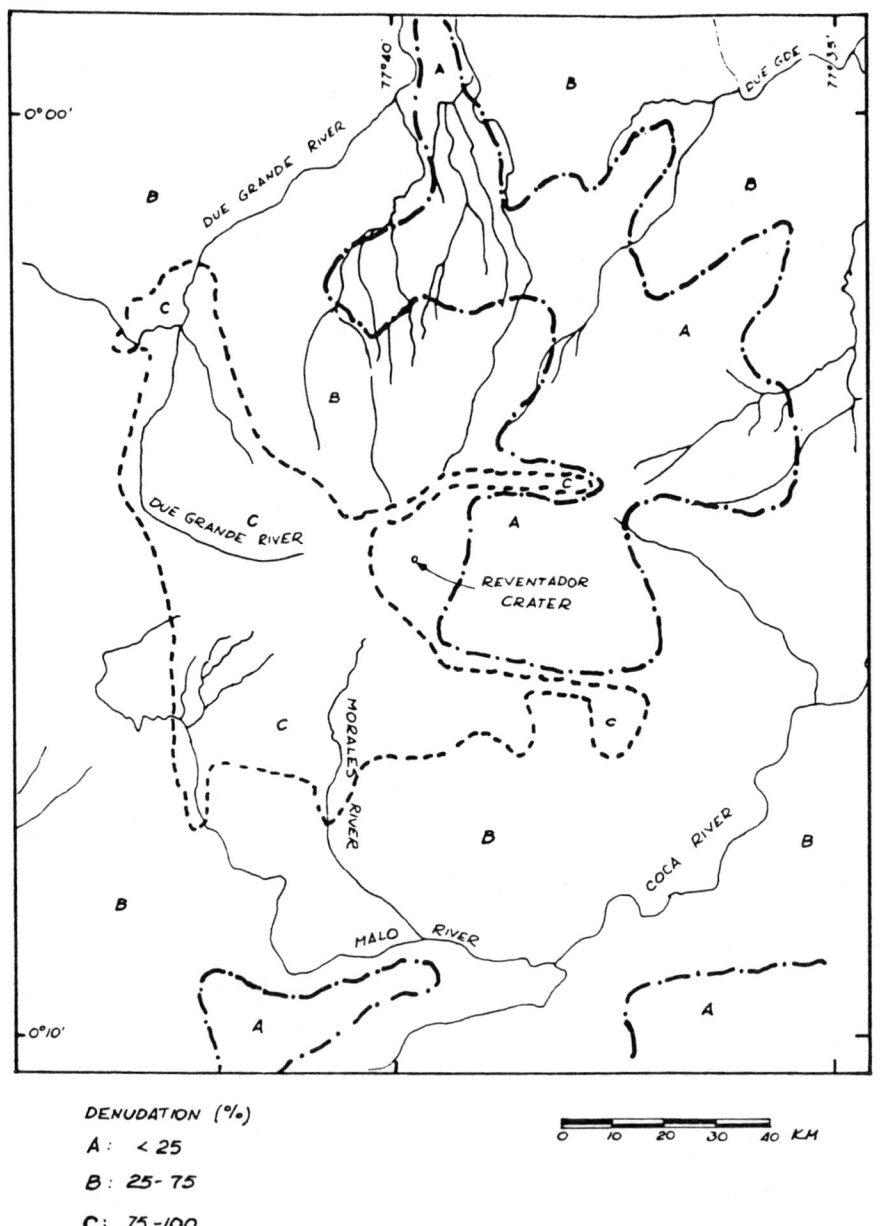

DENUDATION (%)
A: < 25
B: 25-75
C: 75-100

FIGURE 5.10 Map showing percentages of denudation of vegetation due to landsliding in the vicinity of Reventador Volcano, March 5, 1987.

Salado River—variable denudation intensity; range: less than 25 percent to 50 percent; most intense denudation appears to be in first- and second-order gullies of upper reaches of the Salado River, and in third-, fourth-, and lower-order tributaries of lower portions of the Salado River (Figure 5.11).

Quequno, Dué, and Dué Grande rivers—variable denudation; range and local distribution similar to that of the Salado River; however, overall denudation for these areas appears to be somewhat less than for the Salado River drainage basin.

Coca River—between the mouths of the Salado River and the Reventador River, denudation is greatest on the left wall of the Coca River valley; downstream from the Reventador River, denudation is about the same on both sides of the Coca River for comparable elevations. Opposite Reventador Volcano, denudation appears greater on the left side of the Coca valley because the ground surface continues to rise toward the crater, whereas the topography flattens at an elevation of 1,500 and 1,700 m on the right side.

The degree of denudation changes with horizontal curvature of slopes. Airphotos show that there is an increase in denudation on the concave portions of the Coca River slopes and a decrease on convex portions, as seen in plan view.

FIGURE 5.11 Upstream view of the Salado River showing remains (center foreground) of a debris flow that entered the river from the right (NE) valley wall. It is probable that this debris flow briefly dammed the river at this point.

Slope Stability—Preliminary Analysis and Interpretation

This discussion is restricted to the Reventador area because it contains by far the greatest concentration of landslides. The slides were shallow (average depth 2 m) and involved residual soil that became very fluid during failure. Stability thresholds as a function of slope angle are 35 to 40° for the main valley walls of the Coca and Salado rivers and 30 to 35° for the upper portions of the ancient Reventador Volcano cone. Field reconnaissance indicated that relatively few failures occurred on slopes flatter than these values and that most of the slopes with steeper angles did fail.

Evaluating the stability of the slopes in the Reventador area under earthquake loading provides insight into the general shear-strength behavior of these soils in particular and of residual soils in general. The stability of thin residual soils overlying high and steep slopes can be analyzed by an infinite-slope model. Further, the earthquake effects can be evaluated roughly by introducing a pseudostatic horizontal force that models the maximum horizontal acceleration. The analysis is carried out in terms of total stresses because evaluating the pore-water pressures at failure is impossible. Thus, the factor of safety can be defined as

$$FS = \frac{c/\cos \alpha + \gamma H (\cos \alpha - k \sin \alpha) \tan \varnothing}{\gamma H (\sin \alpha + k \cos \alpha)}$$

where
$\varnothing, c$ = consolidated-undrained shear-strength parameters
H = depth of failure surface
γ = unit weight of soil
α = slope angle
k = seismic factor (maximum horizontal acceleration).

Ishihara and Nakamura (1987) used $c = 0.3$ kg/cm^2 and $\varnothing = 30°$ to calculate a static factor of safety for the 45° slope that failed during the earthquakes and destroyed part of the pumping station on the Trans-Ecuadorian oil pipeline at Salado. The cohesion value was obtained from cone-penetration tests and an assumed reduction factor. The value of $\varnothing_{cu}$ (angle of internal friction based on consolidated-undrained conditions) would appear at first glance to be too high for undrained failure conditions (which we presume were considered by Ishihara and Nakamura), but, as discussed below, may be adequate for these particular failures. Landivar et al. (1986), in their general study of the lateritic soils of Ecuador, performed consolidated-undrained isotropic triaxial tests on soils just outside the Reventador zone. They obtained effective-stress strength parameters ranging from 0.05 to 0.08 kg/cm^2 for c' (cohesion on effective-stress or drained basis) and 32 to

34° for ⌀' (angle of internal friction on effective-stress or drained basis). We have plotted the total-stress envelopes for these materials and, as expected, the average $⌀_{cu}$ values for the stress levels tested are in the range of 5 to 10°. The normal stress levels for 2-m-deep slides are not in excess of 0.5 kg/cm^2; thus, these soils were tested at stress levels several times greater than field conditions at Reventador. A review of the literature on structured soils (loess, residual soils, sensitive clays) in general and of residual-soil case histories in particular shows that at very low stress levels the pore pressures at failure are close to zero or slightly negative (Vargas, 1974; Quigley, 1980; and Lum, 1982). As a consequence, the values of $⌀_{cu}$ and ⌀' are very similar, implying that the total-stress envelope has a strong curvature at low stress levels. For our range of interest, 0 to 0.5 kg/cm^2, ⌀ values typically range from 0 to 35°, and c values are generally less than 0.1 kg/cm^2. Therefore, we fitted curved envelopes on the Landivar data and obtained $⌀_{cu}$ values between 30 and 38° and c values between 0.1 and 0.2 kg/cm^2. These values are indeed very similar to the effective-stress parameters for other locations given above. Figure 5.12 shows the variation in the factor of safety as a function of slope angle for static conditions and maximum horizontal accelerations of 150, 250, and 350 gal. A value of α = 30° was used ($⌀_{cu}$ = 30° was also used), and the value of c = 0.14 kg/cm^2 agrees well with the c values obtained from Landivar et al. (1986). It is seen that for the range of validity of Eq. (1), or values between 30 and 60° (which also happens to be the range of slope angles in the field), if instability is assumed at $⌀_{cu}$ = 30° and k = 0.35, instability will begin at α = 40° for k = 0.25 and at α = 45° for k = 0.15. From our field observations, these values appear to be reasonable for the entire Reventador area, where the range of k probably falls within these limits. For comparison, Figure 5.13 shows a similar family of curves, but one in which $⌀_{cu}$ was assumed to be very low (5°), as compared with common assumptions for undrained conditions and a total-stress analysis. Again, c = 0.26 kg/cm^2 was obtained by back calculation assuming FS = 1 and k = 0.35. The use of these strength parameters does not explain failures at angles >45° for k = 0.35 and, more significantly, predicts no slides at lower accelerations.

Furthermore, assuming natural variations in the strength parameters, low $⌀_{cu}$ and high c values do not explain our field observations. Figure 5.14 is a comparative plot in which FS was calculated for k = 0.35 and variations in the strength parameters. When $⌀_{cu}$ = 5° is assumed, a variation of about ±0.05 kg/cm^2, or about 20 percent, from the original 0.14 kg/cm^2 value, gives the dashed curves shown in Figure 5.14. In one case, regardless of the slope angle, none of the slopes would fail, whereas in the other case all slopes would fail. However, even in the areas of maximum denudation, there is good correlation between slope angle and failure. On the other hand, when $⌀_{cu}$ = 30° is assumed, a variation in slope angle of ± 5°, or

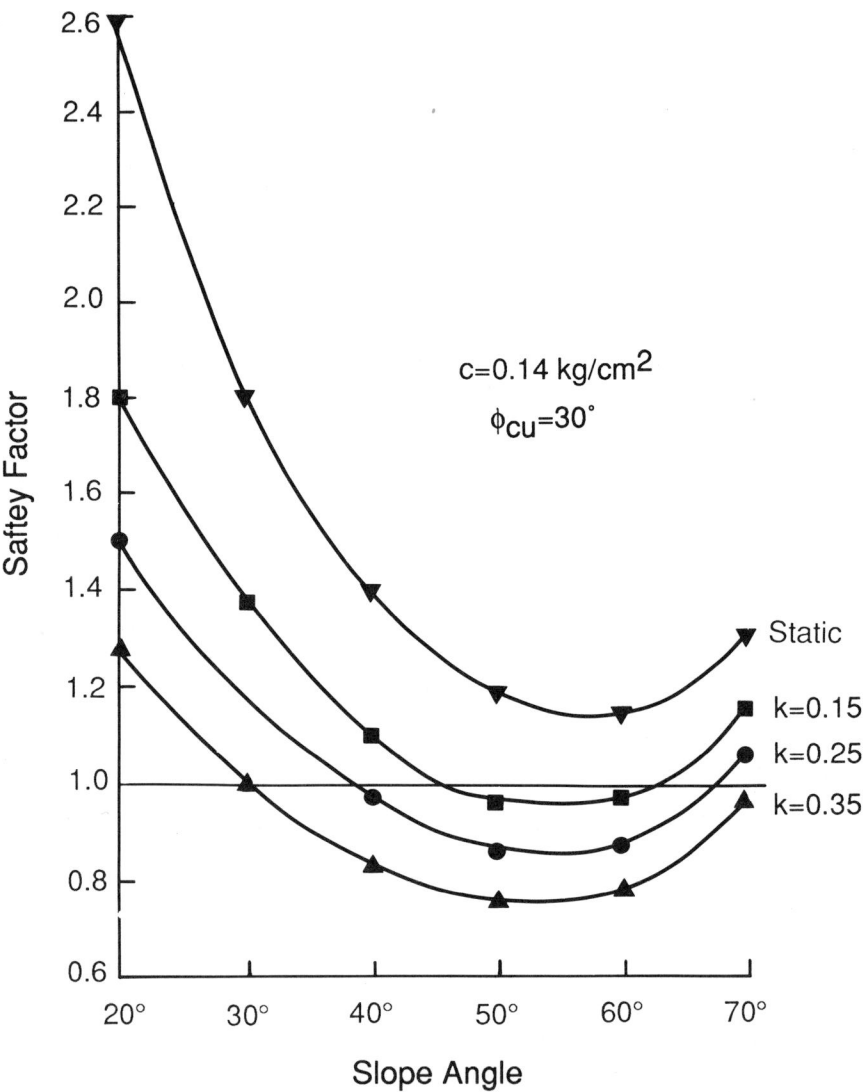

FIGURE 5.12 Pseudostatic and static safety factors vs. slope angle for $\phi_{cu} = 30°$ and $c = 0.14$ kg/cm^2.

about 20 percent from the original 30° value, results in the solid lines in Figure 5.14. These lines indicate that instability for $k = 0.35$ begins on slopes of 35 and 25°, respectively; these slope-angle values still appear reasonable in light of our field observations.

Because these are such shallow slides, the cohesion value that is assumed

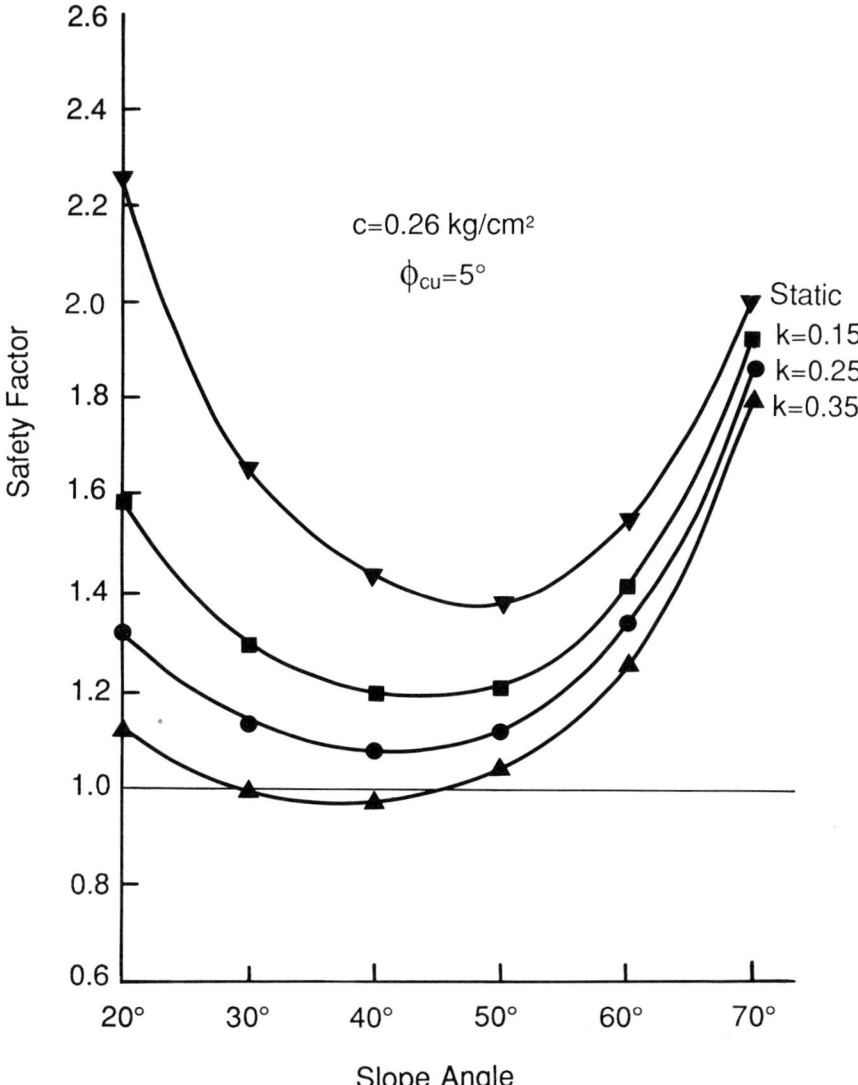

FIGURE 5.13 Pseudostatic and static safety factors vs. slope angle for $\phi_{cu} = 5°$ and $c = 0.26$ kg/cm².

greatly influences the factor of safety even in the case of high ϕ_{cu}. For instance, the value $c = 0.3$ kg/cm² used by Ishihara and Nakamura (1987) can explain the 5-m-deep slope failure at the Salado pumping station with a value of $k = 0.35$, but cannot explain the frequent shallower failures (commonly 1 to 2 m deep) observed nearby.

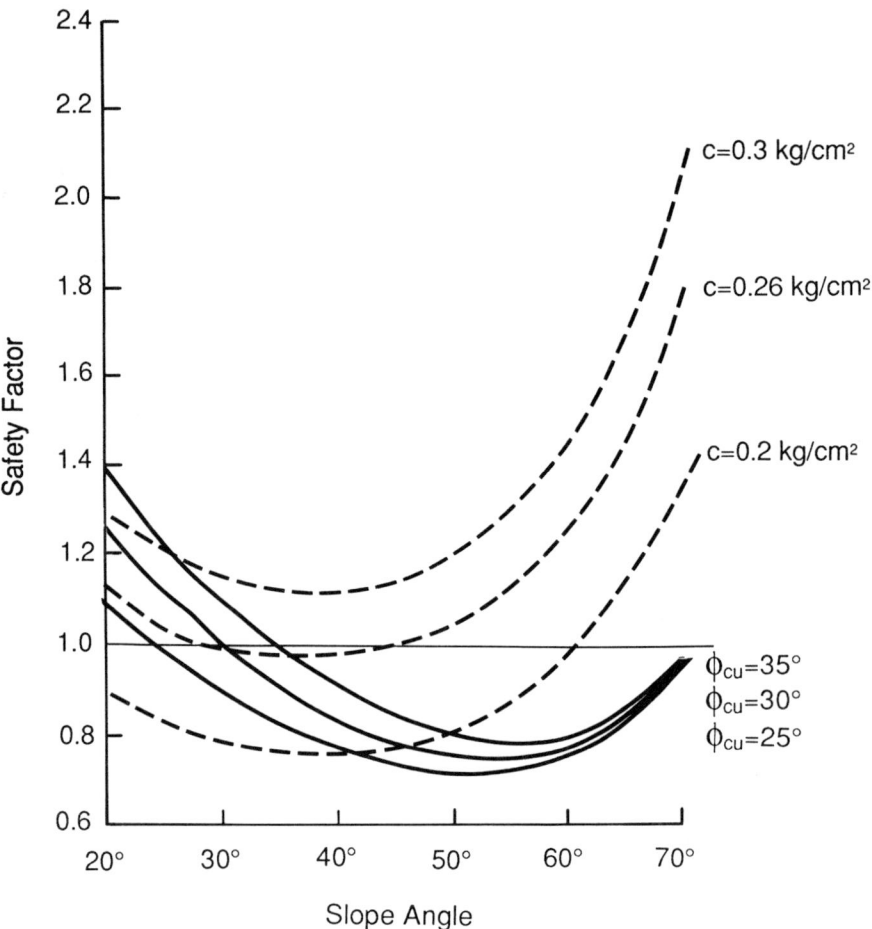

FIGURE 5.14 Comparative plot showing influence of variation in ϕ_{cu} and c on the factor of safety for k = 0.35.

Relation Between Denudation Intensity and Epicenter

Hakuno et al. (1988) questioned the accuracy of the positions of the epicenters that had been estimated by seismologists because the most severe slope failures occurred about 30 km from the reported epicenters. Ishihara and Nakamura (1987) placed the epicenters at 5 to 10 km from the headwaters of the Malo River and 10 to 15 km from the upper Salado River, presumably on the basis of denudation intensity.

Whereas the position of the epicenters may be open to question, we suggest that the increase in denudation near Reventador Volcano may also be caused by other factors, namely relief, moisture conditions, elevation, and soil composition. The areas of near-total denudation on the SW slope of the ancient cone correspond to an area of deep and dense dissection by parallel gullies. Here, almost all the surfaces have slopes greater than 35 to 40°. Near-total denudation in areas of dissection by gullies also has occurred along the walls of the deep canyons of the Malo River (Figure 5.15), Morales Creek, Dué Grande River, and streams to the N of the Dué Grande. In contrast, the slope on the N side of the ancient cone, which is not deeply dissected, is far less affected by landslides, even though it is closer to the epicenters and is less than a couple of kilometers away from the area of almost total denudation. Along the walls of the Coca, slopes that are concave in profile, and therefore have steeper upper slopes and greater density of steep-sided gullies than slopes that are convex, have far greater concentrations of landslides. The same considerations apply to the Salado valley walls. Thus, it would seem that intensity of denudation is strongly controlled by density and depth of dissection, because these factors determine the percentage of slope surface greater than the threshold values.

Another factor controlling denudation may be local differences in degree of saturation of these residual materials. Degree of saturation greatly influ-

FIGURE 5.15 Aerial view of the NE valley wall of the Malo River showing extreme denudation of slopes due to slips/avalanches/flows and of the valley bottom due to debris flows and flooding. Note the vegetation trimline that indicates the maximum height of the debris flow/flood, about 25 m above the current river level.

ences cohesion (Brand, 1982; Ho and Fredlund, 1982), which, as noted previously, affects the pseudostatic factor of safety. The area around the Reventador cone has a wetter microclimate than the rest of the zone (E. Aguilera, personal communication, 1987); the INECEL station near Reventador Volcano has a mean annual precipitation of 6,868 mm.

Last, but not least, the materials near Reventador Volcano, being mostly residual soils formed on pyroclastic materials, may simply be more susceptible to failure.

EVOLUTION OF MASS-WASTING PROCESSES

One of the more striking characteristics of the mass wasting caused by the March 5, 1987, earthquakes was the effectiveness of the transport of materials from the slopes of the lowest-order tributaries to the flood plains of major streams. As mentioned previously, some reaches of the main rivers received as much as 20 m of sediment (measured in the center of the flood plains), most of which was landslide-generated. If we assume a triangular cross section for sediments in the valley bottom, an average width of 600 m for the Coca River floodplain, and a valley length of 20 km between the mouth of the Salado River and San Rafael Falls, we calculate a total volume of 120×10^6 m^3. If we now assume an average depth of landsliding of 2 m (a reasonable value based on field observations), the denuded area was about 60 km^2. The calculated volume compares well with the total volume (110×10^6 m^3) of the mass wastage obtained by Hakuno et al. (1988). This volume attests to the fluidity of the debris flows and the effectiveness of the tributaries in transporting the debris materials. We suggest that two factors may have contributed to this large volume. The first is the nature of the soils involved in the slope failures, and the second is the general morphology of the Reventador area.

In general, the vast majority of the soils involved were tropical residual soils or relatively recent pyroclastic materials of various grain sizes (ash, lapilli, cinders, and pumice). Both of these types of soils have characteristically open structures, and hence high water contents when saturated. The natural water content is typically very close to or higher than the liquid limit. The pyroclastic materials have relatively low liquid limits and plasticity indices. However, the residual soils, being usually at advanced stages of laterization, have high liquid limits (up to 300 percent) and plasticity indices (up to 150 percent). The plasticity of such soils is the result not of high-activity clays, but of the presence of hydrated sesquioxides of Al and Fe in a gel state. In fact, residual laterites have very small or no clay content (Mitchell and Sitar, 1982). If present, clays tend to be of the halloysite type. Both the sesquioxides and halloysite suffer irreversible changes upon drying, and the soils undergo dramatic decreases in plasticity (Mitchell and Sitar, 1982). These hydrated materials, which provide plasticity, behave as cementitious elements that impart physical cohesion to the soils. The open structures are preserved and the result is a brittle, open soil with high capacity to absorb water.

As shown in tests by Hakuno et al. (1988), the soils in the Reventador area fit this characterization well. These authors found natural water contents equal to or higher than the liquid limit, a wide range of plasticities, and absence of clay minerals. Pyroclastic materials, particularly those of fine grain size, also have open structures and relatively low plasticity. These two types of soils exhibit extreme loss of strength if they fail under undrained conditions. The first factor in these dramatic strength losses is the collapse (in the dehydrated residual soils); the second factor is the increase in pore pressures and the attendant decrease in effective stress.

The scenario postulated for the transport of the landslide materials is as follows: (1) Failure at a basal shear surface with an average depth of 2 m, probably at the bottom of the residual soil profile and at the top of the weathered rock. (2) Contractive behavior of the failed soil. Contraction decreases effective stresses until a steady-state strength is reached (Poulos et al., 1985; Ellen and Fleming, 1987; Fleming et al., 1989). Given the highly contracted behavior of the soils in this area, we can surmise that the steady-state strength of these soils is very low. (3) High-velocity flow down steep slopes that are several tens of meters to a few hundred meters long. (4) Movement of debris flows to the thalwegs of tributaries that do not have floodplains. Flow then was channeled and effectively conducted to channels of higher-order streams.

FLOODING OF RIVER VALLEYS IN THE VICINITY OF REVENTADOR VOLCANO

Crespo et al. (1987) roughly estimated that "ground shaking triggered mudslides and rock avalanches near the volcano involving more than 100 million cubic yards [76 million m^3] of soil and rock." Based on airphoto study of denudation of slopes in the vicinity of Reventador Volcano, Hakuno et al. (1988) estimated the total volume of slope failure at about 110 million m^3, but noted that this approximation easily could have been in error by as much as 50 percent. Whichever of these estimates is the more accurate, a large percentage of this huge mass of material combined with water in the Coca and Aguarico Rivers and their tributaries to form thick debris flows that descended these tributaries of the upper Amazon.

Because these debris flows occurred at night, and thus were not easily observed, we are not sure of their character. We do not know whether they acted as true debris flows, or whether they might better be designated as "hyperconcentrated flows" or "mud floods." However, based on (1) depths of sediment deposited in the rivers, (2) "trimlines" on the lower valley walls, which indicated the heights to which the rivers rose when charged with the debris flows, and (3) damage to the Trans-Ecuadorian pipeline and highway due to deposition and erosion, we know that considerable flooding occurred (Figures 5.16A,B, 5.17A,B, and 5.18). Figure 5.19 indicates

FIGURE 5.16A

FIGURES 5.16 Aerial oblique views looking downstream at the confluence of the Salado and Quijos rivers to form the Coca River. (A) 1978 view (photograph by William Savage, Pacific Gas and Electric). (B) April 1987 view illustrating that braided debris-flow and flood deposits (as much as 20 m thick) have covered much of the

estimated depths of flooding and thicknesses of sediments deposited from the floods at several key locations along the Coca and Aguarico Rivers and their tributaries.

Because the stripping of earth materials from the slopes also removed the vegetative cover (mainly trees and brush), these large debris flows undoubtedly were charged with timber debris, similar to the debris flows that de-

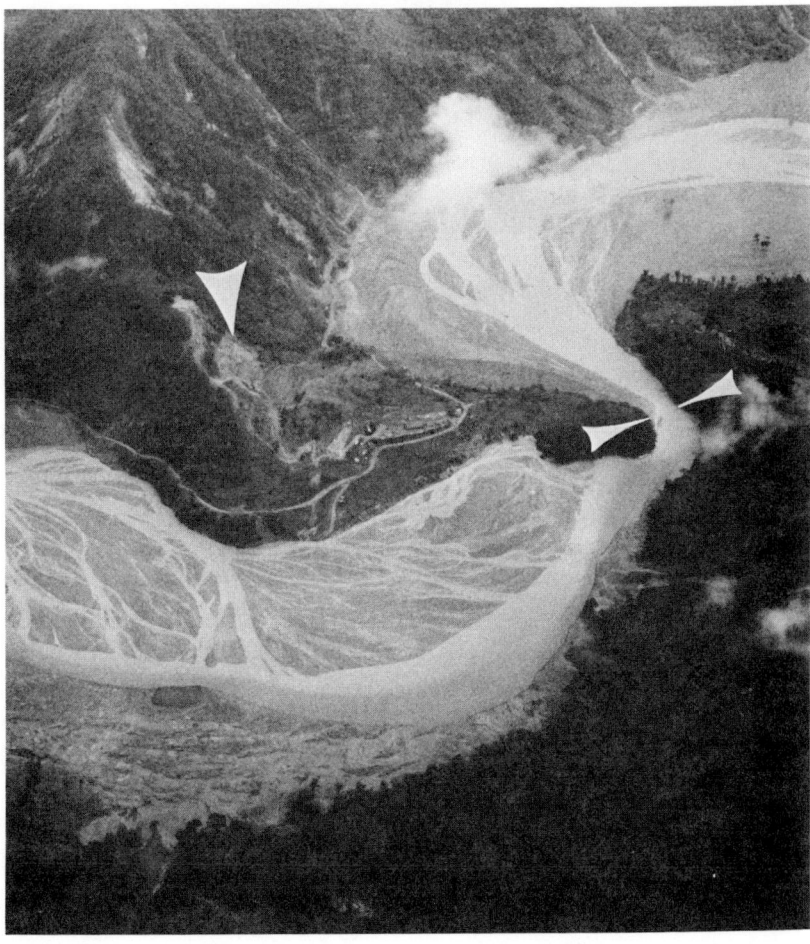

FIGURE 5.16B

floodplain. Bedrock constriction of the Coca River (indicated by two arrows at the right) probably caused short-lived "hydraulic" damming of the river that contributed to upstream flooding and sedimentation. Note landslide (single arrow near center of photo) that badly damaged the Salado pumping plant of the Trans-Ecuadorian oil pipeline.

FIGURE 5.17A

FIGURES 5.17 Looking up the Salado River from the confluence of the Salado and Quijos rivers. (A) 1978 photo showing the Trans-Ecuadorian highway bridge (arrow) across the Salado River. Note the Trans-Ecuadorian pipeline where it snakes down the ridge at the left to its crossing at the mouth of the Salado River.

scended the Toutle River in western Washington State as a result of the 1980 eruption of Mount St. Helens (Schuster, 1983). The addition of this timber debris undoubtedly affected the physical character of the debris flows, and as discussed below, probably impeded their passage through narrow bedrock constrictions in the narrow stream valleys.

As noted by Hakuno et al. (1988), a local resident who lived along the Coca River about 30 km from the epicenters of the earthquakes (most probably between the mouth of the Malo River and San Rafael Falls) reported that the Coca became completely dry soon after the earthquakes, which occurred at about 2100 and 2300 on March 5. Flow began again with a high

FIGURE 5.17B

(Photograph by William Savage, Pacific Gas and Electric.) (B) April 1987 photograph showing extreme sedimentation in the Quijos, Salado, and Coca river channels due to the March 5, 1987, debris flows and flooding. Note absence of the Salado River highway bridge, which had been washed out.

flood at about 0300 on March 6. There is a strong possibility that this interruption of flow of the Coca River was the result of natural damming of the river and/or its tributaries as a result of the earthquakes. We feel that short-lived damming occurred in two ways: (1) "hydraulic" damming, in which stream flow, highly charged with debris, was impeded in passing through narrow bedrock constrictions in the stream channels, and (2) blockage of streams by debris flows issuing into the main stream from its tributaries.

We have noted evidence of "hydraulic" damming at four locations in the Coca River drainage (Figure 5.19): (1) of the Salado River 7 km upstream from its confluence with the Quijos/Coca River (Figure 5.20), (2) of the

FIGURE 5.18 Evidence of March 5, 1987, flooding on the left (N) bank of the Aguarico River near the town of Lumbaqui. The flood has removed much of the vegetative cover from the low terrace in the foreground.

Malo River at its falls about 1 km upstream from its mouth, (3) of the Coca River at the bedrock peninsula that juts southward from the Salado pumping station (located at Salado), and (4) of the Coca River at San Rafael Falls. At each of these locations, the occurrence of damming is indicated immediately upstream by (1) trim lines showing the highest level of flow and (2) extensive deposits of sediment. The sediments probably were deposited in very short-lived lakes that formed as a result of damming at the constrictions before the debris "plugs" were flushed out.

The formation of temporary stream blockages by debris flows issuing from tributaries to dam the main stream has occurred in other parts of the world. For example, Montandon (1933) noted that the Upper Rhine River in Graubunden Canton, Switzerland, was briefly dammed in 1585, 1807, and 1868 by debris flows issuing from the Nolla Torrent. The Xiao River in northern Yunan Province, China, has been dammed briefly seven times in this century by large debris flows that issued from Jiangjia Gully, a major tributary of the Xiao. Each of these short-lived blockages of the Xiao River had heights of about 10 m, and most of them were overtopped and failed within a few days (Li et al., 1986). The Colorado River in the Grand Canyon was briefly dammed in 1966 by a cloudburst-triggered debris flow from Crystal Creek (Webb et al., 1988). In the Reventador area, the only case of such debris-flow damming that we identified was caused by a flow that issued from the Malo River into the Coca River during the night of March 5-6. Apparently, this debris-flow blockage was no more than a few meters high, and it probably was overtopped and breached within an hour or two after forming. However, this low, short-lived dam must have been the main cause of the large amounts of sediment (estimated thickness: 20 m) that were deposited in the channel of the Coca River immediately upstream from the mouth of the Malo River (Figure 5.21). Possibly, another such debris-flow blockage occurred where the Salado River enters the Quijos-Coca River, because a large amount of sediment was deposited at this point (Figures 5.16 and 5.17). The same process probably occurred on a smaller scale at other locations where steep gullies flow into the Salado (Figure 5.11), Malo, and Dué Grande rivers.

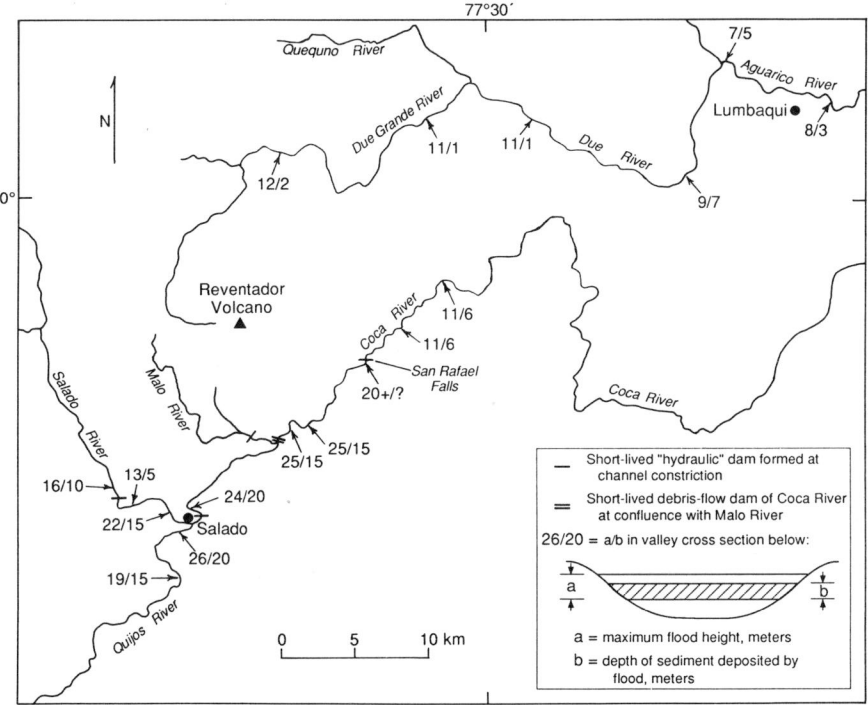

FIGURE 5.19 Estimated depths of flooding and sedimentation in river channels in the Reventador area. Estimates were based on low-level observations from a helicopter.

FIGURE 5.20 Downstream view of the Salado River showing location of bedrock constriction (arrows) that caused short-lived damming of the river. Note trimline in jungle cover along lower left valley wall upstream of the river constriction. This trimline indicates the maximum height (about 10 m above current river level) to which the debris flood rose.

Before the 1987 earthquake, plans had been made by INECEL to construct a 60 to 70-m-high embankment dam at the above-mentioned constriction of the Coca River at Salado. This dam would have formed a reservoir to serve as the source of water for an underground penstock that would reenter the Coca River at a powerhouse downstream from San Rafael Falls (Figure 5.7). Fortunately, construction of this dam had not begun by the time of the 1987 earthquake, and the project was deferred after the quake. However, INECEL is again considering plans for a dam at this site (El Comercio, 1990). The new project would entail a 5-m-high embankment dam at the same site on the Coca River; this low-head dam would store water for the aforementioned penstock and power plant.

REFERENCES

Baldock, J. 1982. National Geological Map of the Republic of Ecuador. Dirección General de Geología y Minas, Quito, scale 1:1,000,000.

FIGURE 5.21 Looking downstream at the confluence of the Malo River (flowing from the lower left) with the Coca River. A debris flow issuing from the Malo River during the night of March 5, 1987, formed a short-lived dam of the Coca River, resulting in deposition of a large amount of sediment in the Coca channel in the right foreground.

Brand, E. W. 1982. Analysis and design in residual soils. American Society of Civil Engineers, Geotechnical Engineering Division. Proceedings of Specialty Conference on Engineering and Construction in Residual and Tropical Soils, Honolulu, January 1982, 89–143.

Crespo, E., and H. E. Stewart. 1987. Stability of cut slopes in Ecuadorian volcaniclastic deposits. Proceedings 8th PanAmerican Conference on Soil Mechanics and Foundation Engineering, Cartagena, Colombia, 3:39–49.

Crespo, E., K. J. Nyman, and T. D. O'Rourke. 1987. Ecuador earthquakes of March 5, 1987. Earthquake Engineering Research Institute, EERI Newsletter, 21(7):1–4.

El Comercio (Quito). June 24, 1990. Coca será una realidad.

Ellen, S. D., and R. W. Fleming. 1987. Mobilization of debris flows from soil slips, San Francisco Bay region, California, in J. E. Costa and G. F. Wieczorek, eds., Debris Flows/Avalanches: Recognition and Mitigation, Geological Society of America Reviews in Engineering Geology 7:31–40.

Fleming, R. W., S. D. Ellen, and A. A. Mitchell. 1989. Transformation of dilative and contractive landslide debris into debris flows—an example from Marin county, California. Engineering Geology 27:201–223.

Hakuno, M., S. Okuna, and M. Michiue. 1988. Study Report of Damage Done by the 1987 Earthquakes in Ecuador. Research Field Group, Natural Disasters and the Ability of the Community to Resist Them, Japan, Research Report on Unexpected Disasters No. B-62-2, 38.

Hall, M. L. 1977. El volcanismo en el Ecuador. Publicación del I.P.G.H., Seccion Nacional del Ecuador, Quito, 120.

Ho, D. Y. F., and D. G. Fredlund. 1982. Increase in strength due to suction for two Hong Kong soils. American Society of Civil Engineers, Geotechnical Engineering Division. Proceedings of the Specialty Conference on Engineering and Construction of Tropical and Residual Soils, Honolulu, 263–295.

Ishihara, K., and S. Nakamura. 1987. Landslides in mountain slopes during the Ecuador earthquake of March 5, 1987. Proceedings of US-Asia Conference on Engineering for Mitigating Natural Hazards Damage, P. Karasudhi, P. Nutalaya, and A. N. L. Chiu, eds., Bangkok, 14–18 December C6-1 to C6–11.

Keefer, D. K. 1984. Landslides caused by earthquakes. Geological Society of America Bulletin 95:406–421.

Landivar, H., C. Luque, O. Ripalda, R. Maruri, and A. Fuentes. 1986. Los suelos lateríticos en el Ecuador. Universidad Catolica de Guayaquil, Ecuador, 39 plus appendices.

Li Tianchi, R. L. Schuster, and Wu Jishan. 1986. Landslide dams in south-central China, in Landslide Dams: Processes, Risk, and Mitigation, R. L. Schuster, ed., American Society of Civil Engineers Geotechnical Special Publication No. 3, 146–162.

Lum, W. B. 1982. Engineering problems in tropical and residual soils in Hawaii. American Society of Civil Engineers, Geotechnical Engineering Division. Proceedings of the Specialty Conference on Engineering and Construction in Tropical and Residual Soils, Honolulu, January 1982, 1–12.

Mitchell, J. K., and N. Sitar. 1982. Engineering properties of tropical residual soils. American Society of Civil Engineers, Geotechnical Engineering Division. Proceedings of the Specialty Conference on Engineering and Construction in Tropical and Residual Soils, Honolulu, January 1982, 30–57.

Montandon, F. 1933. Chronologie des grands éboulements alpins, du début de l'ère chrétienne à nos jours, in Matériaux pour l'Etude des Calamités, Société de Géographie Genève 32:271–340.

Poulos, S. J., G. Castro, and J. W. France. 1985. Liquefaction evaluation procedure. Journal of Geotechnical Engineering, American Society of Civil Engineers 111(6):772–792.

Quigley, R. M. 1980. Geology, mineralogy and geochemistry of Canadian soft soils: a geotechnical perspective. Canadian Geotechnical Journal 17:261–285.

Schuster, R. L. 1983. Engineering aspects of the 1980 Mount St. Helens eruptions. Bulletin of the Association of Engineering Geologists 20(2):125–143.

Vargas, M. (1974) Engineering properties of residual soils from south-central region of Brazil. Proceedings of the Second International Conference of the International Association of Engineering Geology, Sao Paulo, Brazil, 1.

Webb, R. H., P. T. Pringle, S. L. Reneau, and G. R. Rink. 1988. Monument Creek debris flow, 1984: Implications for formation of rapids on the Colorado River in Grand Canyon National Park. Geology 16:50–54.

6

Effects on Lifelines

E. Crespo,* Cornell University, Ithaca, New York
T. D. O'Rourke,* Cornell University, Ithaca, New York
K. J. Nyman,* Cornell University, Ithaca, New York

GENERAL OBSERVATIONS

From March 19 to April 3, 1987, reconnaissance personnel were sent from Cornell University to Ecuador to evaluate the effects on lifelines of the March 5, 1987, earthquakes. An additional visit was made on August 17-27 to collect data concerning the effects of landslides and flooding on key lifeline facilities and to observe reconstruction efforts in areas most severely influenced by the earthquakes. Although the primary focus of the reconnaissance missions was the Trans-Ecuadorian pipeline, considerable effort was spent on gathering information about other lifelines. Visits were made to the City of Quito and towns in the provinces of Pichincha, Imbabura, and Napo (Figure 4.1). Interviews were conducted with government and municipal workers as well as with local people.

Damage to lifelines was severe in areas near the earthquake epicenters. Specifically, there was major damage to the Trans-Ecuadorian (crude oil) and Poliducto (propane) pipelines, as well as to the principal highway linking Quito and Lago Agrio (Figure 6.1), the main town of the oil-producing region of Ecuador. Damage to lifelines in other areas was relatively light and is summarized briefly under the following subheadings.

Water Distribution Systems

Officials of EMAP, the municipal agency responsible for the water supply to Quito, were interviewed. They reported that no damage was sustained to the pipeline network, pump facilities, and reservoirs servicing

---

* School of Civil and Environmental Engineering

83

Quito. The only damage to underground water mains that was observed was in the town of Baeza (Figure 6.1), which is approximately 60 km from the epicenters. This observation was made at one location, where a 150-mm-diameter PVC pipeline was being repaired. The full extent of pipeline damage in Baeza could not be determined; however, there was considerable damage to one- and two-story structures, primarily of reinforced masonry construction.

Electric Power Systems

Officials with EEQ, the municipal electricity authority for Quito, were contacted. They reported that electricity had been interrupted to all neighborhoods of Quito for more than 2 hr after the second earthquake (M=6.9). This interruption was attributed to the functioning of emergency breaker switches. Inspection of electrical equipment after the earthquakes indicated that some maintenance and minor repair were needed.

No damage to hydroelectric facilities was reported by INECEL, the national electricity institute. A small, privately owned hydroelectric plant in the town of Papallacta (Figure 6.1, Table 6.1), approximately 50 km from the epicenters, was visited. No damage was observed in the power plant or its piping facilities.

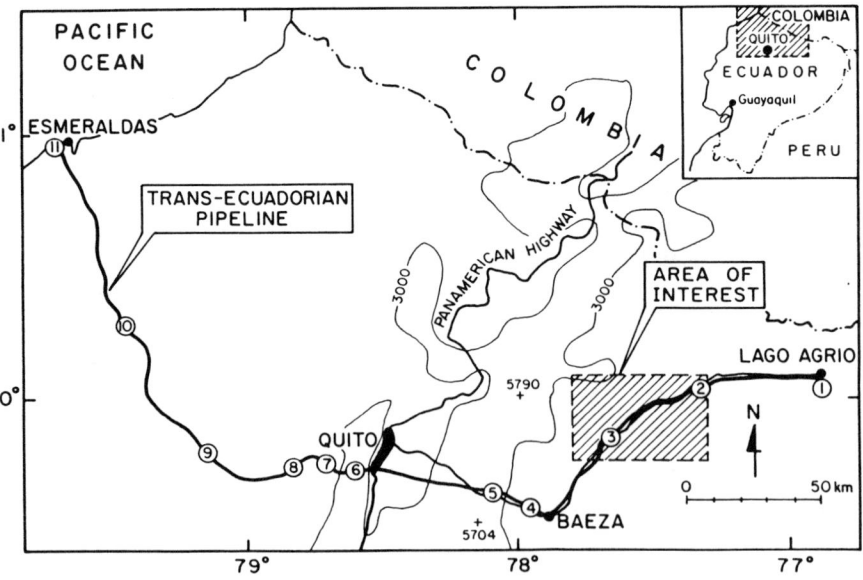

FIGURE 6.1 Route and station locations of the Trans-Ecuadorian pipeline.

TABLE 6.1 List of Pump and Pressure-Reduction Stations on the Trans-Ecuadorian Pipeline

Number	Name	Elevation (m)
1	Lago Agrio Pump Station	296
2	Lumbaquí Pump Station	842
3	Salado Pump Station	1,268
4	Baeza Pump Station	2,001
5	Papallacta Pump Station	3,007
6	San Juan Pressure-Reducing Station	3,496
7	Chiriboga Pressure-Reducing Station	1,997
8	La Palma Pressure-Reducing Station	1,612
9	Santo Domingo Pressure-Reducing Station	566
10	Quininde Pump Station	96
11	Balao Marine Terminal	0

Pan-American Highway

The Pan-American Highway is the major transportation facility connecting the towns and cities of the Andean region. Landslides in volcaniclastic deposits were caused by the earthquakes and interrupted traffic between the cities of Quito and Cayambe (Figure 1.1). The majority of these landslides originated in the faces of steep roadcuts and involved relatively shallow raveling and toppling of cemented sands, ash, and weak rock.

Transportation between Quito and the town of El Quinche (20 km ENE of Quito) was interrupted by a relatively large landslide 1 week after the earthquakes. This landslide occurred in volcaniclastic deposits at a steep roadcut. Approximately 24 hr was required to clear the road at this location.

Natural Gas Supply

Natural gas in Ecuador is used primarily for cooking and is supplied in pressurized canisters and tanks of propane. There are no natural-gas-pipeline distribution networks in Ecuador. There are several transmission pipelines, which are used to convey propane, and one of these, known as the Poliducto pipeline, was damaged by the earthquakes. A more detailed description of this line and associated damage is given under the next two major headings.

Although earthquake damage to lifelines was light throughout most of the country, damage near the epicenters was so severe and widespread that it had a major economic impact on the country, with repercussions felt worldwide. Approximately 40 km of the Trans-Ecuadorian pipeline had to be reconstructed, making this the single largest pipeline failure in history. Because of the scale and importance of the damage, this chapter concentrates on the characteristics and postearthquake observations related to the Trans-Ecuadorian pipeline, with supplemental descriptions of the Poliducto natural gas pipeline and adjacent highway.

CHARACTERISTICS OF THE TRANS-ECUADORIAN AND POLIDUCTO PIPELINES

The Trans-Ecuadorian pipeline (crude oil) is composed primarily of 660-mm-diameter line pipe and associated pump and pressure-reducing stations. The pipeline was commissioned in 1972. It is composed of X-60 grade steel with wall thickness ranging from 9.5 to 20.6 mm. The pipeline is the main crude oil transportation facility in the country, conveying virtually all oil from the eastern oil fields to a marine terminal port near Esmeraldas on the Pacific Ocean (Figure 6.1). The pipeline operates at a maximum internal pressure of 9.7 MPa and moves from 250,000 to 300,000 barrels of oil a day.

The Poliducto pipeline is composed of 150-mm-diameter line pipe and associated compressor station pressure-regulation equipment. Its wall thickness is 7.1 mm. The line was built after the Trans-Ecuadorian pipeline was commissioned, and closely follows the right of way for the crude oil line. The Poliducto pipeline is a multiproduct facility. It conveys different types of hydrocarbons at different times, including propane gas. The line extends from Lago Agrio in the eastern oil field to Quito.

Figure 6.1 shows a plan view of the Trans-Ecuadorian pipeline, with pump stations at locations 1 through 5 and 10, and pressure-reducing stations at locations 6 through 9. The endpoint of the pipeline at the Balao marine terminal is situated near Esmeraldas at location 11. The pipeline conveys oil from an initial elevation of 296 m at the Lago Agrio pump station to a maximum elevation of 4,060 m at a mountain pass E of Quito to the Balao marine terminal at sea level. The elevations relative to sea level of the various pump and pressure-reducing stations are summarized in Table 6.1. For purposes of clarity, the Poliducto pipeline is not shown, although it should be recognized that this facility follows the same right-of-way corridor from Lago Agrio to Quito as illustrated in the figure.

The total length of the Trans-Ecuadorian pipeline is 498 km from Lago Agrio to Esmeraldas. Approximately 200 km, from Lago Agrio to a location roughly halfway between Quito and the Papallacta pump station, is constructed primarily above ground. Along this part of the route, the line pipe is supported every 12 m on either an "H" pile support frame or concrete-pedestal foundation saddle. The support frame consists of two 150-mm-diameter pipe piles driven 6 m into the ground, connected by a pipe cross beam welded to the driven piles. The pipeline rests on the cross

EFFECTS ON LIFELINES

beam, which is fitted with a curved center plate saddle. A concrete pedestal consists of a 1-m-square foundation footing with a cradle keyed vertically into the footing. The 50-mm-tall vertical concrete section is fabricated to a semicircular shape at the top, which acts as a saddle to support the pipe. The above ground supports do not have special features to restrict pipe movement either laterally or longitudinally. On steep terrains, however, longitudinal anchors are used periodically to prevent slippage of the pipe downhill through the supports.

The remaining 298 km of the pipeline from E of Quito to Esmeraldas is buried. One section of the line, from location 8 to 9 as depicted in Figure 6.1, is composed of 510-mm-diameter pipe. All other sections are composed of 660-mm-diameter pipe.

The area of most severe disruption is shown by means of the hachured zone in Figure 6.1. It was in this general region that landslides and floods triggered by the earthquake damaged the line pipe and the Salado pump station (location 3).

PIPELINE DAMAGE

Figure 6.2 shows an expanded view of the hachured zone identified in Figure 6.1. This general region includes the pipeline route from its crossing

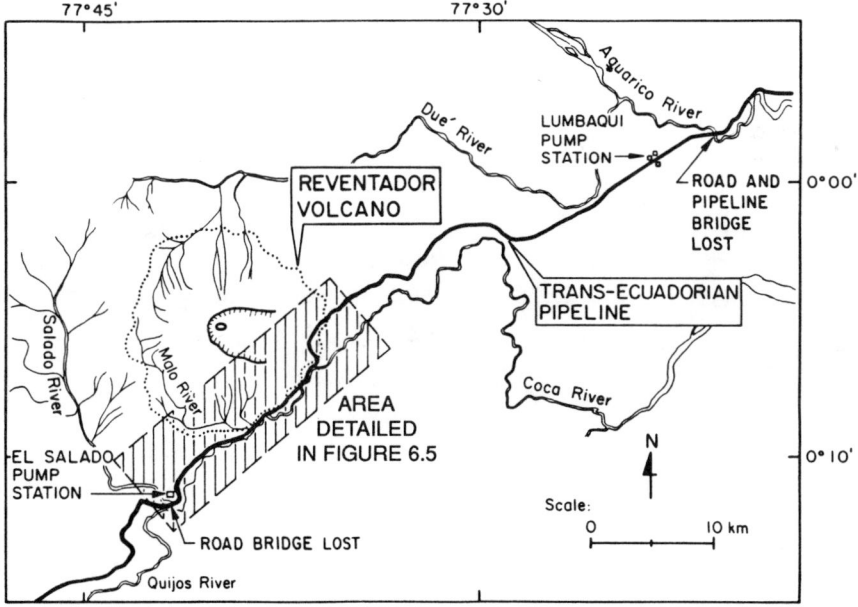

FIGURE 6.2 Area of major pipeline damage.

of the Aguarico River to its crossing of the Salado River. The pipeline alignment follows a SW course along the Coca River basin. Between the Aguarico and Salado Rivers, the pipeline attains a maximum elevation of 1,700 m, after which it descends approximately 500 m to the Coca River floodplain and then follows the N bank of the river to the Salado pump station.

The Coca River is deeply entrenched and borders the southeastern foothills of the active volcano, Reventador. Reventador Volcano is a large stratovolcano characterized by a nested cone in a horseshoe-shaped rim that opens toward the E. The composite cone is the expression of a recent phase of volcanism. In 1977, the volcano produced lava flows that extended to within 2 km of the pipeline.

The two earthquakes of March 5, 1987, occurred after a month of heavy rain, during which 600 mm of precipitation was measured at a nearby rain gage. Strong ground shaking triggered rock and soil slips, avalanches, and debris flows. The flows and resulting floods drained northward into the Dué River and southward and eastward into the Salado and Coca Rivers.

Flooding along the Dué River entered the Aguarico River and destroyed the pipeline and highway bridge across the Aguarico approximately 50 km NE of the Salado pump station. Flood waters reached an elevation of 5 m above the bridge deck. Figure 6.3 shows a segment of the buried pipeline,

FIGURE 6.3 Trans-Ecuadorian pipeline severed by flood waters, E bank of the Aguarico River.

FIGURE 6.4 Trans-Ecuadorian pipeline on the W bank of the Aguarico River pulled southward by flood waters. Note the remnant pier and abutment of the bridge.

which was severed on the W side of the river at its junction between the bridge and concrete abutment. Figure 6.4 provides a view of the E side of the river where the pipeline had entered the bridge abutment as an aboveground structure. W of this location, the pipeline was pulled longitudinally and displaced off its supports for a distance of 4 km.

The highway bridge at the Salado River crossing was completely destroyed. The Trans-Ecuadorian pipeline at this location had been constructed by dredging a 3- to 4-m-deep trench in the river bottom, into which the line was placed and backfilled. The pipeline was encased in concrete to offset buoyancy effects. The pipeline was undamaged at the Salado crossing, and this segment of the line was incorporated without repair into the rebuilt pipeline system.

Figure 6.5 provides an expanded view of the area (hachured zone in Figure 6.2) of greatest pipeline damage. Approximately 12 km of the pipeline was destroyed along the banks of the Coca River, from just E of the Salado pump station to a location roughly 12 km E of the confluence of the Salado and Coca rivers. Along this section, damage was generated by landslides and debris flows in the residual soils and igneous parent rock of the hills flanking the N side of the Salado River. Most damage, however,

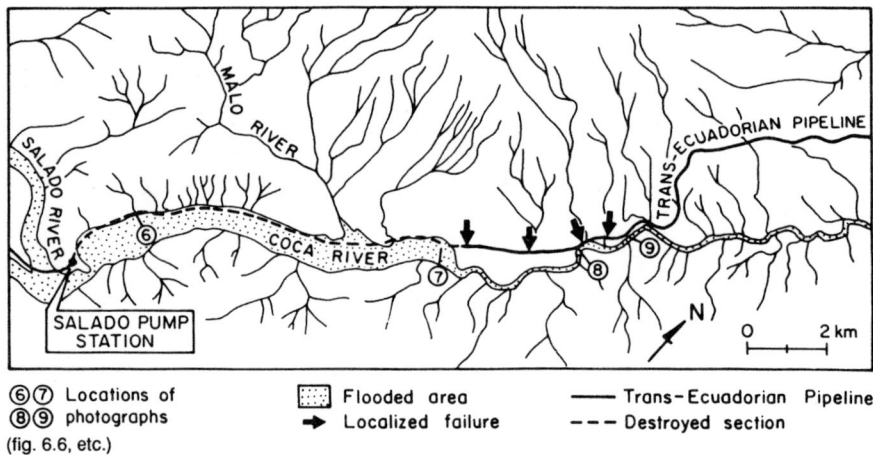

FIGURE 6.5 Area of flooding, landslides, and debris flows.

was caused by flooding of the Coca River, which is shown by the stipled area in the figure. Flooding encroached upon the pipeline alignment, resulting in severe scouring and removal of the entire pipeline right of way. Two sections of the pipeline, each approximately 1 to 2 km long, were left intact along this portion of the river. These sections were recovered during reconstruction and used on a provisional basis to establish flow in the rebuilt line.

As described previously, the 150-mm-diameter Poliducto Pipeline followed approximately the same route as the Trans-Ecuadorian pipeline. It was destroyed and damaged at approximately the same locations as the Trans-Ecuadorian pipeline.

Approximately 5 km of the Trans-Ecuadorian pipeline, immediately E of the section described above, was damaged at locations where landslides and debris flows intersected the line. The black arrows in Figure 6.5 show the points of localized damage, where the pipeline was severed. Many landslides and debris flows in this section originated in the younger volcaniclastic deposits surrounding Reventador Volcano at elevations between 1,400 and 2,000 m.

The pipeline was severed by landslides and debris flows at two other locations approximately 7 and 10 km E of the area depicted in Figure 6.3. At one location, it was ruptured by a landslide in volcanic debris at a steep slope immediately adjacent to the line. At the other location, it was severed by a debris flow that originated as a landslide in a steep slope of thinly bedded shale.

Figures 6.6 through 6.9 show photographs taken during helicopter reconnaissance approximately 3 weeks after the earthquakes. The locations of the photographs are identified as 6 through 9 in Figure 6.5, each corresponding to its respective figure number.

FIGURE 6.6 Pipeline broken by a debris flow that has eroded 6 m below the pipeline right of way.

FIGURE 6.7 Flooding along the Coca River. Note the broken and deformed pipeline at the left side of the photograph.

FIGURE 6.8 Trans-Ecuadorian and Poliducto (white) pipelines ruptured by debris flow and deep erosion along the Coca River.

FIGURE 6.9 Broad fan of a very large debris flow with the Trans-Ecuadorian pipeline deformed in the direction of debris movement.

SALADO PUMP STATION DAMAGE

The Salado pump station is located near the confluence of the Salado and Coca rivers on terrace deposits at an elevation approximately 25 m above river level. At this location, alluvial sands, gravel, and cobbles extend to a depth of approximately 200 to 300 m below the floodplain of the Coca River (Almeida and Cruz, 1986).

A plan view of the station is shown in Figure 6.10. Severe damage in the station was caused by a landslide that occurred in weathered granodiorite and terrace deposits at an elevation roughly 80 m above the station. The debris from this slide traveled approximately 240 m to the E, where it ruptured the principal oil tank of the station and buried the main gate valve. The location of the landslide and the outline of the debris flow are shown in the inset diagram in Figure 6.10.

As shown in Figure 6.10, the Salado pump station is composed of several structures, the most important of which include the main gate valve, control building, generators, pump house, water tank, crude oil tank, communication equipment, and personnel housing. Each of these is labeled in the figure. The operational and physical characteristics of these structures and a brief description of their postearthquake condition is given under the headings that follow.

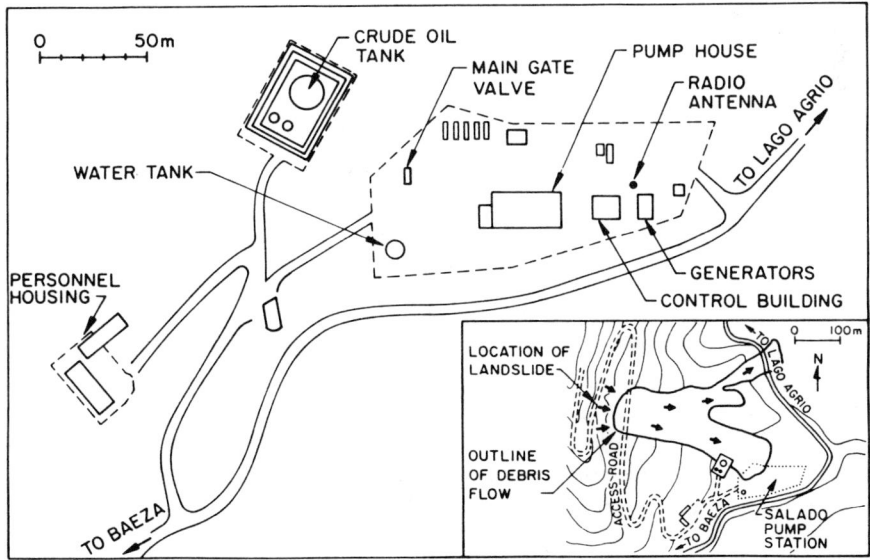

FIGURE 6.10 Plan view of the Salado pump station.

Main Gate Valve

The main gate valve controls the flow from the station and, therefore, controls downstream pressure in the line. It allows for isolation of the line in the event of difficulty. Debris from the landslide completely buried this facility. Because the remote control system of the station was lost, burial by debris prevented manual operation of the valve. As a consequence, an unknown but significant volume of oil in the line E of the pump station was lost by flow from ruptured sections W of the pump station.

Control Building

The control building was a one-story reinforced-concrete structure on a continuous concrete slab foundation. It contained the control equipment necessary for proper operation of the pump station. Visible damage to this structure consisted of a 12-mm-wide crack along the length of the control-room floor slab. The control-room operators indicated that the switching panels had been damaged by the earthquakes. Electrical power was lost throughout the station immediately after the second main shock (M=6.9). Backup equipment and emergency generators failed to operate.

Generators

Two diesel-powered generators are used as the main source of electric power for the station. Approximately 12 mm of differential settlement and over 25 mm of horizontal displacement were observed at the concrete slab foundation of one of the generators. During site reconnaissance, approximately 3 weeks after the earthquakes, this generator was still under repair. In addition, one of the elevated diesel fuel tanks for the generators overturned as a result of the earthquakes.

Pump House

The pump house contained five diesel engine pumps, each supported on isolated concrete blocks and underlying 300-mm-diameter steel pipe piles. Soil settlement relative to the pile-supported blocks ranged from approximately 35 to 100 mm. This settlement apparently contributed to differential movement between the pump motors and the heat exchangers, which were supported on shallow spread foundations. This differential settlement caused distress in the rotating connections. Figure 6.11 shows the pump house during reconstruction of the station. In the photograph, excavation along the side of the structure has exposed the pile-supported block foundation.

EFFECTS ON LIFELINES

FIGURE 6.11 Pump house at the Salado pump station. Note soil settlement relative to pile-supported foundation blocks.

Communications Equipment

The main radio antenna buckled and became inoperable. This antenna was used to transmit continuous operations information and was the principal means of emergency communication.

Water Tank

The main 10-m-diameter water tank was 5.5 m high; it was constructed of approximately 7-mm-thick steel plate. Its purpose was to supply water for fire fighting. The 2,000-barrel tank was full at the time of the earthquake. As shown in Figure 6.12, wrinkling occurred near the top of the tank, apparently in response to sloshing. A large circumferential bulge, or "elephant's foot" buckle, developed at the juncture between the tank and its concrete ring wall (Figure 6.13). The main pipe outlet from the tank pulled free of a compression coupling, thereby disconnecting the tank from the station piping network.

Crude Oil Tank

The 15-m-diameter oil tank, which provides surge-pressure relief and oil storage, was about half full during the earthquake. The tank was con-

FIGURE 6.12 Damaged water-storage tank at the Salado pump station.

FIGURE 6.13 Circumferential buckling and pipe pullout of the water tank at the Salado pump station.

FIGURE 6.14 Oil tank destroyed by landslide debris at the Salado pump station.

structed of 12-mm-thick steel plate and was approximately 10 m high. Debris from the landslide hit the tank, causing it to collapse and to spill roughly 4,500 barrels of oil over the station. Figure 6.14 shows a photo of the collapsed oil tank and two smaller fuel tanks that were destroyed by the landslide and debris flow. Excavation and inspection during reconstruction revealed that oil from the tank had seeped through the ground to the Coca River over 100 m away, infiltrating subsurface soils for a depth of approximately 10 m. An oil fire erupted near the location of the ruptured tank during reconstruction of the station about 2 months after the earthquakes.

Personnel Housing

The barracks and mess halls were one-story reinforced concrete structures. Severe cosmetic damage was observed throughout the housing. Some structural damage to the reinforced concrete beams in the mess hall was also observed.

TRANS-ECUADORIAN HIGHWAY FROM BAEZA TO LAGO AGRIO

The road parallel to the pipelines (Figures 5.1, 6.1, 6.2) is the main transportation artery from Quito to the eastern oil field. Flooding destroyed the highway bridges at the Salado and Aguarico rivers as well as large portions of the road between the Salado and Malo rivers. More than 6 months after the earthquakes, this road was still disrupted, and the eastern region still experienced shortages of supplies because of the loss of this transportation lifeline. The only alternative means of transportation to Lago Agrio were by plane or by a combination of road and river travel. (The Salado River bridge was replaced in 1988 by a "Bailey" bridge, which currently is in use.)

ECONOMIC CONSEQUENCES

Loss of the Trans-Ecuadorian pipeline deprived Ecuador of 60 percent of its export revenue. As a consequence, the loss of this single lifeline had a dramatic effect on the country's economy. Assuming that the pipeline could transport roughly 250,000 barrels daily at an average price of $19 per barrel, the total lost revenue from March 5 to the first provisional use of the reconstructed line on August 18, 1987, was approximately $790 million. The cost of pipeline reconstruction is estimated at approximately $50 million.

The price of West Texas intermediate crude oil is often used as an index of the world oil price. News of the earthquakes and associated loss of the Trans-Ecuadorian pipeline was followed by a 6.25 percent increase in the price of West Texas crude oil over the 4 trading days immediately following the earthquakes. Although oil prices had been climbing at the time of the earthquakes, market analysts claim that the news of Ecuador's suspension of oil exports encouraged trading at escalating prices. As a consequence, the economic effects of the lifeline failure were not confined to a single country, but were felt worldwide through market speculation.

SUMMARY

Earthquake damage to the Trans-Ecuadorian pipeline represents the largest single pipeline loss in history. Even though the March 5, 1987, earthquakes occurred in a remote region, their consequences in terms of lifeline failure had significant national and international ramifications.

Seismic shaking had only a limited effect on the line pipe, whereas permanent ground deformations had a severe and extensive influence. Landslides and debris flows caused most of the pipeline damage and contributed to virtually all ruptures and permanent deformations of the line. Damage from landslides and debris flows was caused directly by failed soil or rock material that intersected the line, or indirectly by rivers swollen with debris that flooded the line and eroded the pipeline right of way.

Pump station damage at Salado was extensive. Nearly all pump station systems failed to operate properly after the earthquakes. The most severe structural damage was caused by a landslide that destroyed the crude oil tank at Salado, spilling thousands of barrels of oil across the station. Permanent differential settlements and horizontal displacements were observed at most building foundations and were responsible in part for malfunctioning of control equipment, connections between pumps and heat exchangers, and generators. Seismic shaking damage was especially severe at the water storage tank and radio communications antenna.

ACKNOWLEDGMENTS

This study was made possible thanks to the logistical support of CEPE/Texaco, Inc. Special thanks are extended to several employees of CEPE/Texaco, Inc.: Dr. Juan Quevedo, General Manager; Mr. Bob Paulsell, Assistant Superintendent; Mr. Bill Spear, Chaco Camp Superintendent; and Mr. Jerry Isacks, Quito Superintendent. Mr. Gustavo Freile of Harbert Engineers provided critical assistance during both reconnaissance visits. Mr. Ivan Nunes of INECEL provided valuable information. We thank Messrs. "Swamp" Smith and Bill Spencer of Will Bros. Construction Company for their assistance and hospitality. We express our gratitude to Capt. Francisco Hidalgo of Ecuador's Army Aviation. Special thanks and recognition are given to Prof. John M. Bird of Cornell University for his assistance and insight during the second reconnaissance trip.

REFERENCE

Almeida, E. and M. Cruz. 1986. Estudio geológico del Volcán Reventador, INECEL, Proyecto Geotérmico, Quito.

7

Local-level Economic and Social Consequences

P. A. Bolton, Battelle Institute, Seattle, Washington

INTRODUCTION

While not a major focus of the North American media, the earthquakes that occurred in northeastern Ecuador on the night of March 5, 1987, were of considerable significance within Ecuador. At the national level, the country's already deteriorating economy suffered a major blow when Ecuadorian oil production was disrupted by earthquake-related damage to the Trans-Ecuadorian pipeline. According to the United Nations Economic Commission for Latin America and the Caribbean (ECLAC, 1987), the Ecuadorian oil fields had accounted for about 60 percent of the nation's export earnings. Thus, Ecuador's ability to meet its internal operating costs and to make interest payments on its foreign debt were severely impaired. Within a week after the earthquakes, the national government instituted several extreme economic measures, including suspension of the external debt payment to private banks, increased fuel prices, a national austerity plan, and a price freeze on selected essential goods.

The earthquake's social consequences and accompanying demands for response and recovery assistance were relatively unusual with respect to their variety and geographic scope. Damage to the Trans-Ecuadorian pipeline resulted in both a major reconstruction cost and a major loss of national revenue. The damage to housing was spread over a large area and distributed mostly among households of limited resources. The loss of a major transportation route created problems for thousands of people who otherwise suffered no direct damage.

At the most basic level, the tremors were particularly frightening to the populations most directly affected. No earthquakes of a similar magnitude had occurred in the affected area of the country for more than 30 years. The two major tremors, 3 hr apart, left in their wake not only a variety of types of damage and social consequences, but considerable further hazard in the form of thousands of weakened buildings and the threat of additional landslides. Fortunately, within the next few months there were no substantial aftershocks, although about 20 of the thousands of aftershocks that occurred were perceptible to human beings.

This chapter describes the local-level consequences of the earthquakes that were noted during brief visits to three of the affected provinces. The observations reported here are mainly based on visits to several of the affected communities to view the damage, and on conversations with a variety of local officials, recovery program representatives, and local residents encountered along the way. A few documents and newspapers also were reviewed. The information from the communities was gathered approximately 16 weeks after the earthquakes, and thus refers mostly to consequences experienced and actions taken in the disaster-stricken areas up to the middle of June 1987.

The communities discussed here are located in regions of Ecuador known generally as the *Oriente* and the *Sierra* (Figure 2.1). The Oriente extends eastward from the flanks of the Andes and forms part of the Amazon Basin. Damage to the communities in the Andean region, known as the Sierra, occurred mainly in the relatively densely populated central valley N of the capital city, Quito.

As described in earlier chapters, the mass-wasting phenomena that resulted from the earthquakes occurred in the mountainous western fringes of the Oriente in the vicinity of Reventador Volcano (Figure 1.1). The communities visited in this region (Figures 1.1, 1.2, 4.3) included Lago Agrio at the northern edge of the affected area, Baeza at the southern end, and some smaller villages between Baeza and the bridge site where the Salado River joins the Coca River. These communities are located in Napo Province.

In the Sierra, where damage was mostly limited to specific types of structures affected by ground shaking, exploratory visits were made to parts of the capital city of Quito, to the town of Tabacundo, the village of Olmedo, and the city of Ibarra and its environs (Figure 4.3). These communities lie in northern Pichincha Province and in southern Imbabura Province (Figure 4.1). Carchi Province (Figure 4.1), N of Imbabura Province, also was considered part of the disaster area, but was not visited.

IMMEDIATE CONSEQUENCES AND EMERGENCY-RESPONSE ISSUES

There was little major damage to structures in Quito, although observers reported some rather spectacular fireballs that accompanied the consequences to the power system of the first strong ground movement. Within a few hours, the electricity and phone service were back to normal. Later damage assessments revealed minor cracking in various buildings around the city, and somewhat more serious damage to several colonial-era churches and cathedrals in the oldest section of Quito, as well as to some of the older houses.

The earthquakes were felt with sufficient intensity in Quito to alert national and international agencies to the possibility of serious consequences and to the need to begin an immediate assessment of the situation. The various environmental clues led to some degree of uncertainty in the first few hours. For example, preliminary damage assessments had to be reexamined once the second and more damaging quake (M_s=6.9) occurred 3 hr later. Also, there was initial concern, because of the location of the epicenter and reports of landsliding, as to whether a volcanic eruption had been involved, or might be expected. Secondly, even before dawn, because of the evidence of landslides that resulted in flooding along the Coca and Aguarico rivers and in a broken oil pipeline along the Coca, attention of emergency responders and assessment teams was initially focused on the greatly affected but sparsely populated area immediately E of Reventador Volcano (Figure 1.2). The Ecuadorian military detachment in Lago Agrio was first on the scene on Friday, March 6, beginning its reconnaissance and search-and-rescue missions at dawn.

On Saturday, March 7, assessments were made by U.S. experts. Also by Saturday, the extensiveness of the damage to houses in the Sierra had become evident. For the most part in the Sierra, damaged houses had not collapsed immediately, and there was no loss of life or need for search-and-rescue efforts.

The major effects were determined to be of three general types:

• Direct effects of the landslides, debris flows, and flooding to the infrastructure in the area, including damage to roads and bridges and in particular the Trans-Ecuadorian oil pipeline and the parallel Poliducto gas line. This damage in turn had secondary effects on both the local and national economies.

• Direct effects from the ground shaking on housing and some public buildings in communities N of Quito, and also to some extent in the Oriente.

• Indirect effects on the population of Napo Province that no longer had land access to the rest of the country, as a result of the only road from the town of Lago Agrio to the Sierra region and the capital city of Quito being impassable.

This chapter provides a brief review of the emergency-response-related issues and of the most significant aspects of the emergency period as they differed in the Oriente and the Sierra. It then describes emerging long-term impacts observed in June 1987, followed by sections on the economic and social impacts and recovery activities observed in the Oriente and the Sierra.

THE EMERGENCY PERIOD IN THE ORIENTE

Beginning with the effects on Napo Province, the most evident direct physical damage from the earthquake was on the flanks of Reventador Volcano and on the floodplains of the drainage systems in this mountainous area. Virtually all of the loss of life associated with the event occurred in Napo Province. The most common estimate of the number of deaths related to the earthquakes is about 1,000. Those who lost their lives were caught by the landslides, or were swept away in the rivers swollen by the debris flows of saturated soil, rock debris, and vegetation from the steep volcanic slopes. These victims typically were residents of plantations or small settlements in the hills and on the floodplains in the area between Baeza and Lumbaquí (Figure 7.1).

In general, the area has only recently been settled by farmers who came there as part of the national agrarian reform and colonization program. Previously, the area was inhabited by various indigenous groups that to a

FIGURE 7.1 Typical preearthquake dwelling along the Trans-Ecuadorian pipeline and highway near Reventador Volcano. (1978 photograph by S. D. Schwarz.)

great extent have been pushed farther into the jungle as land is provided to the new settlers. Estimates of the numbers of settlers that were killed or missing as a result of the earthquake vary considerably because there are no reliable data on how many people were living in the area affected by the landslides, and it is assumed that many bodies have not been recovered from the rivers. No specific information was found on whether or not indigenous settlements were included among those destroyed.

Those who were fortunate enough not to be caught in the landslides and debris flows were stranded if they were located on the N side of the Coca River between the bridge at the confluence of the Salado and Coca rivers and the bridge over the Aguarico River. Most of those who were stranded were evacuated by helicopter in the first day or two after the earthquakes, because it was believed that the area was too hazardous for them to remain.

These evacuations and the search-and-rescue operations were carried out by the Ecuadorian Air Force and Special Forces. Provisional shelter was found in convents, schools, and private homes until tent camps could be established in Lago Agrio and in the villages between the Salado River bridge and Baeza. It generally is estimated that around 4,000 to 5,000 people were evacuated. About 500 to 1,000 people eventually were taken to camps in Quito, and about 1,000 stayed in the camps in the Oriente. It is likely that the majority of the evacuees moved in with relatives or friends near the disaster area or in other parts of Ecuador. Some families stayed in Lumbaquí and other nearby settlements, but had to have supplies airlifted across the Aguarico River to them. Eventually this airlift was replaced by canoes and, for a short time, a footbridge (which subsequently washed out). Airlifts for supplies were coordinated by an emergency operations committee that met daily in Lago Agrio, and were carried out by the Ecuadorian Air Force. Lago Agrio itself did not suffer flooding or earthquake damage.

In the area most heavily damaged by the landslides, much vegetation was lost from the mountainsides, leaving the area even more vulnerable to future landsliding. Plantations, grazing land, and other agricultural developments, as well as livestock, also were destroyed by landslides, debris flows, and flooding. Many vehicles were lost or damaged, and others were stranded in the area until the road could be reopened. This caused hardship for survivors who depended on the vehicles for their livelihood.

The sediment in the rivers from the landslides and debris flows did considerable damage to fisheries for great distances downstream. Although about 100,000 barrels of oil spilled into the river when the pipeline was broken, any environmental effect it may have had was overwhelmed by the effects of the sediment and other debris in the water. The destruction of the fish population undoubtedly had negative consequences, in particular to indigenous groups engaged in subsistence fishing. There also were reports of sediment and other contamination in the rivers causing short-term health problems and making the water unusable until the rivers cleared.

Some tourism was said to have been affected for a short time by the debris load in the rivers, since one of the attractions in eastern Napo Province is boat excursions along the jungle rivers. However, most of this activity is centered in towns that are farther E or S; thus, the economic effects, if any, were not felt in Lago Agrio.

In the small towns S of the area most damaged by landslides, some dwellings were damaged by the groundshaking. In particular, houses constructed of concrete blocks suffered the greatest damage, since most of them had been poorly constructed. To a great extent, these were the homes of people who had prospered enough to move out of the more traditional wooden houses. A few wooden houses also suffered some degree of distortion, or had damage to their concrete foundations. Baeza, the largest town S of the landslide area, presented the most significant example of extensive damage to concrete block homes (Figure 7.2), as well as to some wooden structures.

Although no long-term camp for evacuees from the landslide zone was established in Baeza, many of the town's own residents were dislocated from their damaged or collapsed homes and had to find other lodging or

FIGURE 7.2 Damaged concrete-block dwelling in Baeza, Napo Province.

FIGURE 7.3 Tents for refugees from the landslide/flood zone, placed in the town square of Quijos.

make temporary shelters near their damaged homes. Several of the smaller towns became locations for tent camps that housed evacuees from the landslide area who had nowhere else to go, or who preferred to remain near their landholdings (Figure 7.3). An Italian emergency medical response team established a field hospital in Baeza immediately after the disaster, which they later donated to the community.

THE EMERGENCY PERIOD IN THE SIERRA

The damage caused in the communities N of Quito was characteristic of what most typically happens in an earthquake event like that of March 5. Damage patterns were related to the distribution of buildings with certain characteristics of construction design and quality, with some variation depending on the type of soil and the orientation of the structures in relation to the epicenter.

The most extensive damage in the Sierra occurred to houses constructed of adobe bricks or *tapia* (mud poured into molds and then sun-dried in place to form walls), and particularly to those constructed according to traditional building practices in the rural areas. Consequently, the bulk of the households directly affected were those in the lowest income class. These households typically were located in the more marginal barrios of the larger towns or in the predominantly Indian villages and rural settlements. To some extent, other types of houses and public buildings also were damaged, particularly those constructed in the 19th century or with improper construction practices.

For the most part, the houses did not collapse outright, although there may have been failure of a wall or two, or the collapse of a portion of a roof. No deaths were reported as a result of building collapse. However, the houses that were damaged were greatly weakened, and many collapsed later or were purposely demolished.

Many schools were damaged or destroyed because they had been constructed in the same manner as the houses. A much smaller proportion of community health-center facilities sustained major damage, since these were more likely to be newer buildings than were the schools. In some instances, the health centers served as public shelters in the first week or so.

Although tremors had been felt on occasion in the Sierra in past years, none has jolted the area with the intensity of those occurring on March 5. We were told that residents of the villages exhibited extreme fear. Many engaged in fervent prayer, because they viewed the event as punishment for sinful behavior. However, these typically were secondhand reports, making it difficult to determine whether they were accurate descriptions of the villagers' behavior or whether the behavior of a few individuals was being attributed to the rural population as a whole in order to underscore the local impact of the event.

Health-center patients in Imbabura Province were screened for symptoms of mental health problems 10 to 12 weeks after the earthquakes. Although the data were not yet analyzed as of June 1987, the preliminary impression of researchers at the Ministerio de Salud Pública, División Nacional de Salud Mental (Ministry of Health, National Division of Mental Health) was that about 25 percent of the victims interviewed exhibited emotional problems, to a great degree related to the earthquakes.

After the earthquakes, many people chose to sleep outside their homes for several weeks, even though the weather turned rainy and cold the day after the tremors. This rational adaptation is common in areas hit by damaging earthquakes. As in other earthquake situations, the vulnerability of many of the dwellings was apparent, even without deaths having occurred, and mild aftershocks could be felt for many weeks, creating doubt about whether or not the worst had yet happened. Even people whose homes were not damaged were reported to have slept outside their houses for some time.

More than 1,000 tents were eventually supplied to families in the affected area in the Sierra. Some families slept in public buildings, such as health centers, until other shelter could be prepared. Many other families improvised shelters near their damaged dwellings, using plastic sheeting provided to victims by assistance agencies.

Some tent camps for refugees were established; in other instances, tents or shelters made of plastic sheeting were placed on private lots adjacent to damaged dwellings. Emergency-period activities and distribution of relief goods most commonly were coordinated by the provincial staff of the Defensa Civil de Ecuador (Ecuadorian Civil Defense), although various observers noted that the organization was not as yet very skilled in such activities. Within a few days after the disaster, there was some distribution of clothing and food, but these activities were of short duration, and done in many instances without any systematic attempts to identify who were victims and who were not. The aid distribution was handled differently in different locations; for example, in some places the distribution involved only Civil Defense, at others a priest or minister. It was apparent that most of the disaster-assistance activities during the emergency period were conducted in the more urbanized areas.

Three refugee camps were established on the periphery of Ibarra, the largest city in Imbabura Province, providing temporary housing for residents whose homes were destroyed. These camps apparently were established and maintained under the direction of a local physician, who donated his time to this effort and to tending to the health needs of victims for the first month after the disaster. The tents came from outside Ecuador and were distributed by Civil Defense to the local Red Cross chapter to establish the camp. This physician, working with Red Cross emergency medical assistants (*socorristas*) he was training, planned and established the camp and the procedures for feeding the refugees and maintaining sanitary conditions. The National Guard handled discipline and security. The physician remarked that he knew there were books available from international organizations to guide in the establishment of refugee camps, but he had great difficulty in obtaining them quickly enough to be of any use. He eventually obtained some by going to Quito and paying for them himself, but this was after most of the initial work had been done to establish the camp.

Little mention was made of injuries suffered in the earthquakes. However, the physician in Ibarra noted that respiratory diseases were common among the persons who lived in the tent camp or who had been living outside their damaged houses in the rain and cold.

Mention was made by some of disruption of tourism in the Sierra. Locals observed that the number of tourists visiting Indian markets near Cayambe and Ibarra diminished somewhat after the earthquakes, but only for a short time. Also, one person noted that some people believed that the tourism bureau had made efforts to minimize reports of damage from the region, so as to minimize economic consequences to resorts and markets. The concern was expressed that by minimizing reports of damage, the chance for the area to receive available and necessary disaster assistance was in turn affected. This dilemma has been noted in other disasters and warrants further study in future disasters.

EMERGING LONG-TERM IMPACTS

When considering the longer-term impacts that seemed to be emerging in the communities visited, it is important to recognize the overall economic context in which these were occurring. Although the opening of the Amazonian oil fields many years before had given an important boost to the economy of the country, Ecuador had at the same time become very dependent on oil revenues. Thus, when oil prices fell in 1986, the annual economic growth of the country dropped to below 2 percent, a large fiscal deficit was incurred, and the trade surplus fell (ECLAC, 1987, p. 2).

Including the anticipated 6-month curtailment in oil production while the Trans-Ecuadorian pipeline was being repaired, the total cost to the country was estimated soon after the disaster to be about $1 billion (ECLAC, 1987, p. 26). At about the same time as the March 5 earthquakes and resulting landslides, widespread flooding (unrelated to the earthquakes) occurred in the vicinity of Guayaquil (Figure 1.1); considerable infrastructure was damaged by all of these catastrophes, placing additional demands on the national budget. The reconstruction needs created by the disasters only added to Ecuador's already serious economic difficulties, and rather specific financial strategies were needed to try to deal with the situation.

Under the circumstances, recovery projects designed to help generate local currency are important to consider. National and international strategies for doing this have not been examined in this quick reconnaissance study but, where implemented, warrant evaluation as to their effectiveness and transferability into other postdisaster settings.

To the people in the affected communities, who had already begun to feel the consequences to their pocketbooks of the national economic problems before the earthquake, there was some skepticism that the economic problems and national austerity measures attributed to the earthquake did indeed originate with it. It is difficult to say which of the economic problems evident at the community level after the disaster might have been experienced even without an earthquake, as the national government grappled with its longer-term problems. The natural disasters certainly exacerbated the situation.

The earthquake damage had affected for the most part persons who already were experiencing difficult economic circumstances. Much of the damage occurred in rural areas that had very limited infrastructure and were lacking in basic services even before the earthquake. Thus, the mingling of reconstruction issues and development issues could be expected. For example, one demonstration of dissatisfaction with the progress of disaster recovery was covered in the Quito newspapers in June (Hoy, June 16, 1987; El Comercio, June 17, 1987). The reports described a contingent of local government officials and residents marching to Quito from the capital of Napo Province to ask for additional reconstruction assistance; however, the request was more general than for just things that would allow reconstruction of the disaster-stricken communities to predisaster conditions. Here again, no in-depth analysis was attempted of this aspect of the recovery process, but the relationship between reconstruction and development is a topic of interest for designing disaster assistance in developing countries (Bates, 1982; Cuny, 1983).

Increased prices for staples were part of the overall economic changes that individuals reported in all parts of the disaster area. Gasoline was rationed for a short time after the earthquakes. The price of gasoline increased by about 80 percent at the time, and was still at that level in June 1987. This in turn affected bus fares and the costs of transportation for agricultural produce and other goods. The price of cooking fuel used in urban areas also increased. People reported that the price of some, but not all, food items increased. There also was a shortage of sugar beginning in June, but this was attributed to actions on the part of the sugar producers in order to raise prices.

Many people indicated skepticism about the reasons for the price increases and said that the earthquake was being used as a pretext for the government to institute economic measures it had planned to take anyway. However, the fact that the nation could not meet even its own internal petroleum needs after the pipeline breakage cannot be ignored in relation to increases in fuel prices.

RECOVERY PROGRAMS AND IMPACTS IN THE ORIENTE

By mid-June 1987, little had been accomplished in the way of reconstruction in Baeza and the other communities along the road toward the Salado River bridge. The people in Baeza who had been dislocated from their homes were for the most part either living in shelters on their property or sharing homes with relatives. Some people had apparently left the area, but we were told that many people in Baeza had government jobs and thus their incomes had not been affected, even though they may have had housing problems. It is likely that the restaurant and hotel business had slowed somewhat with the reduced traffic along the Baeza-Lago Agrio highway.

The tent hospital that the Italian medical team had donated was being operated by the Ecuadorian Ministry of Health, in order to provide free health care that had not previously been available in the area. There was a private Catholic hospital in Baeza as well. The tent hospital was staffed with two nurses who lived in Baeza and two doctors who commuted from Quito. The plan was to eventually move the health facility into a permanent building. One of the nurses noted that the main malady they were treating was scabies, which was rampant in the households that had been living in tents for the last 3 months.

A housing loan program offered by the Banco Ecuatoriano de la Vivienda (Ecuadorian Housing Bank) was just being started in the Oriente as of the middle of June 1987. The plan was to have four technical representatives of the bank. These representatives would visit residents, offer loans, and give technical advice on construction. In addition, there would be four social workers, who would check on the families' abilities to qualify for the loans, and would help them find solutions for any financial problems that might arise later. (This loan program is discussed further in the next section.)

In the town of El Chaco (Figures 1.1 and 4.3), we were told that Civil Defense officials were that same day giving out a bag of cement to each family having a damaged house, but few other details about the program were known. Several damaged buildings were evident in the community, but little visible construction or repair work was going on. However, it had been raining daily, which might account for the lack of such activity in the communities.

In talking to a few people in the towns of El Chaco and Quijos (on the Quijos River, 4 km downstream from El Chaco), it was evident that, in the communities with evacuee tent camps, there were three distinctions used for disaster victims: (1) people from other towns who lived in the camps, (2) people who lived in the town whose houses had been badly damaged or destroyed, and (3) people who had not suffered damage. Those in the last group had not received any type of disaster assistance, but often were impacted in some indirect way. Impacts reported included reduced income as a result of a voluntary reduction in agricultural production because it was not possible to transport the agricultural products to market, a reduction in income from food sales during the time that disaster victims were receiving free food, and reduced business in general because the destruction of the Trans-Ecuadorian highway east of the Salado River reduced the road traffic to that point, and many people had left the area to try to find employment opportunities elsewhere.

It was reported that the crews that had formerly operated the pumping stations along the pipeline had mostly left the area. However, there was a construction camp in the town of Quijos for the workers who were replacing the bridge across the Coca River at Lumbaquí (Figure 1.2). One person observed that these workers were paid very good salaries, but described the consequences of this in negative terms because of the workers' disorderly behavior when off duty. However, there undoubtedly were some local benefits from having these jobs and salaries in the area.

Other comments were made by local residents about the relations between various groups, for example, between evacuees and residents, Catholics and Evangelists, victims and nonvictims, and people with money and people without. These comments suggest that further in-depth study might indicate that these communities have fairly low social cohesion. The different social character of people of the Oriente compared with those of the Sierra was mentioned by more than one observer.

In general, aside from the original indigenous inhabitants of the Oriente, who were not encountered during the visit to the disaster area, the populations in the towns and on the plantations were fairly recently arrived, and typically had left some other part of the country in order to seek economic opportunity as colonists to this area. Thus, these residents were not likely as yet to have long-term attachment to the land nor to have extensive and strong social ties to others in the area. One local storekeeper remarked that, right after the earthquake, the people had helped each other, but now things were "back to normal."

Another notable aspect of the earthquakes' effects on the Oriente was the impacts in Napo Province created by the loss of the Salado and Aguarico river bridges. First, the inaccessibility of the land along the approximately 67-km stretch of road between these two bridges created problems for the surviving farmers and plantation owners who had been evacuated from the area. Besides the washed-out bridges, the roads in the area between the bridges were blocked by landslides and by flood damage in many places, and the threat of further landsliding was high, making the area both difficult and very risky for resettlement at that time, even if it could be served by air or boat. Secondly, a large proportion of the 75,000 inhabitants of Napo Province were effectively cut off from the rest of Ecuador by the damage to the road between Baeza and Lago Agrio.

The effects on the inhabitants of the town of Lago Agrio (Figure 1.1) and of the areas to the N and E in Napo Province were for the most part indirect. There was little significant direct damage from the groundshaking of the earthquake in the eastern part of Napo Province. However, agricultural producers in the region suffered significant economic impacts as a result of not being able to transport their crops to market. For example, it was estimated that the postearthquake production losses from abandonment of land or lack of access to markets amounted to about $7 million (ECLAC, 1987, pp. 18-19). This estimate was based on the assumption that land access would be reestablished by the end of June, which it was not. The main link with the rest of the country as of June 1987 was by air, a means of transportation that is too expensive for most agricultural products.

Without income from agriculture in the area, retailers and services in turn suffered from reduced business volume. People familiar with Lago Agrio prior to the earthquakes observed that the number of retail stalls on the main street had greatly declined, and also believed that people were leaving the area increasingly to look for work elsewhere.

An inquiry into the impacts on employment due to the drastic reduction in oil production revealed that the workers at the CEPE-Texaco installation near Lago Agrio figured little in the economy of the town, both before and after the earthquake. The CEPE-Texaco "camp" operated much like an offshore drilling rig, with workers sleeping in dormitories and eating in the company dining hall for several days at a time, and then being flown by the company back to their homes in Quito when the shift changed. The Texaco workers for the most part were shifted from production to maintenance work while the pipeline was being repaired.

With the exception of the indigenous groups, the settlers in the area are dependent on receiving food staples and gasoline from outside Napo. Also, major medical services and some kinds of business services can be obtained only in the larger cities elsewhere in the country. The airlink that existed in June 1987 was used mainly for passengers and for bringing in these staples. The air service was provided by the national airlines, and created considerable uncertainty for persons trying to use it to leave the area. There were only a few flights a day, despite the high demand for the service. The fare could be paid in advance, but seats could not be reserved; thus, hundreds of people stood in long lines for every flight out of the area, and they often could not get on a flight even after waiting a day or two.

One source indicated that products could be transported to and from eastern Napo region by motorized canoe up the Napo River, continuing to Lago Agrio by bus from the town of Coca (also known as Puerto Francisco de Orellana) (Figure 1.1). This apparently was the preferred means for bringing gasoline to the area, and could be used for taking products out. Another source suggested that considerable extortion was involved in getting transportation, air or water, for whatever products did leave the area.

The temporary lack of a road between Lago Agrio and Ecuador's other commercial centers made it difficult for many to maintain themselves economically while this condition existed. More importantly, there also was no specific information available to the inhabitants of the region as to when the road would be repaired. In fact, an alternative route (see Chapter 8) had been proposed and construction started, which if it were completed, would solve the problem of a transportation route for agricultural products, but also would have long-term implications for persons who owned land along the original route.

The question of whether the original road would be repaired or a new route established was not solved at the time of our visit to the disaster area.

Certainly, the existing route was located in a very hazardous area, which had been made even more hazardous by the earthquake-caused landsliding. Although the road and the Trans-Ecuadorian oil pipeline parallel each other much of the way between Baeza and Lago Agrio, and some type of access road is necessary for pipeline maintenance, it was noted that only a minimal road was necessary to service the pipeline. Thus, road construction efforts undertaken by Texaco and CEPE (Corporación Estatal Petrolera Ecuatoriana), the national petroleum company that operates the pipeline, in order to obtain access to the pipeline, would not necessarily result in a road of the same size and quality as existed before the earthquakes.

Continued uncertainty about the timing and location of a road to Lago Agrio made it impossible for the inhabitants of this isolated region to make informed decisions about the best course of action to protect their own personal economic interests. For the most part, about the only option for most was to leave the area in order to survive economically. This also was not a particularly promising option if it meant abandoning one's parcel of land.

Another type of issue existed with respect to the question of access to landholdings along the Baeza-Lago Agrio road and access to Lago Agrio. First, some settlers (about 80 families) who had been farming in the landslide/flood zone had expressed interest in relocation. Among the reasons for their wanting to do this was that the former land was now ruined, or that they could not wait any longer to have land to work, or that they considered the area too hazardous to return to. However, there were difficulties in obtaining other land in the Oriente that could be made available to them. INCRAE (Instituto de Colonización y Reforma Agraria Ecuatoriana), the Ecuadorian colonization and agrarian reform agency, indicated that there were major problems to be solved, both for the families that were willing to stay in the disaster area to farm and for those who wanted to be relocated.

In order to obtain new land in northeastern Ecuador, progress needs to be made in agrarian reform. Whether families are returned to their pre-earthquake plantations or resettled onto new lands, resources are needed to provide the necessary infrastructure of roads, bridges, schools, health centers, and so forth. For either solution to be accomplished, several national-level agencies will have to work in concert to bring it about in a timely manner. Thus, at least 200 families who had farmed the land between the Salado and Aguarico River bridges remained in limbo about their futures 4 months after the earthquakes, many living without a permanent dwelling or income.

RECOVERY PROGRAMS AND IMPACTS IN THE SIERRA

In the Sierra, many families with badly damaged or demolished houses constructed simple shelters on their lots, and in June 1987 were preparing to rebuild their houses. Others reportedly had doubled up with relatives for the short term until they could rebuild. Some of the housing construction activity was undertaken with technical supervision and some was not. Because there was no mechanism for condemning dangerous buildings, there also was concern that many families continued to live in houses that were even more vulnerable to future earthquakes than they had been before the earthquakes. The dilemma of reconstructing communities in a timely manner versus taking the time and resources to provide safer housing and other facilities has been observed in many other disasters (Haas and others, 1977; Bates, 1982).

The reconstruction period provides one of the best opportunities for upgrading housing and making it safer, because interest and awareness are high. In the damaged villages of northern Pichincha Province (Figure 4.1), the provincial office of Civil Defense coordinated housing reconstruction with the Ecuadorian Housing Bank and other groups. They developed a plan for the reconstruction that focused mainly on helping the poorest victims.

A map of the projects in Pichincha Province showed three different types of projects: (1) communities in which housing reconstruction was supported by Civil Defense, (2) communities supported from international assistance organizations ("outside the [national] plan"), and (3) communities that had been designated for a project, but for which there were no funds as yet. The projects to be implemented jointly by Civil Defense and either the housing bank or private contractors were slated to start in October 1987. For these projects, construction materials were to be given to the homeowners who were to provide the labor, with technical assistance from Civil Defense or the housing bank. Necessary machinery was to be loaned by the municipal government. At the time of the author's reconnaissance visit, it was not possible to say to what extent any of the planned housing programs would be implemented or in which villages this would occur.

The town of Tabacundo (Figure 4.3) provided one example of a community in which many stages of housing reconstruction could be seen at one time. The temporary housing being used ranged from crude, makeshift shacks, to tents from foreign donors, to small houses of wood slabs. Many lots had been completely cleared of the damaged structures, and construction was beginning (Figure 7.4). Some residents were rebuilding with their own resources, while about 120 households had received metal frames, donated by a German organization, to provide support for walls made of brick infill. Other people were building more traditional adobe buildings. The community was given a machine for compacting the adobe bricks to make them more seismic-resistant, and we were told that technical assistance was available in the community from Ecuadorian agencies. The people could request certain building materials from Civil Defense; these materials were delivered by trucks provided by the municipal government.

FIGURE 7.4 Temporary dwelling of plastic sheeting used as a residence while family builds a new home using a metal frame, Tabacundo, Pichincha Province.

In most small Andean communities, the population of Indian descent typically uses the traditional social practice of the *minga*, or ad hoc work group, to complete community projects. A specific project is selected and a work group, sometimes involving most of the people in the village, is formed to complete it. In Tabacundo, the minga tradition was being used in the reconstruction of housing. For example, a minga work group was observed making adobe bricks using a new compacting device (Figure 7.5).

Various interviews indicated that there was some controversy about what approach to reconstruction has the fewest drawbacks. Governmental officials claimed that the Indians, even though they are primarily farmers, had sufficient time to also work on the reconstruction of housing in their communities. Furthermore, the use of the residents to provide the construction labor makes the projects less costly and supposedly gives the residents a stronger sense of ownership for reconstruction activities. Other observers felt that the construction work cut into the farming work and that this might lead to problems in the future if food production was inadequate. Another argument was given by some that it was *not* the best idea to have the villagers do the construction, since houses not properly designed and constructed will be likely to suffer damage during future earthquakes.

FIGURE 7.5 *Minga* (traditional community work group) making mud bricks using a brick-compacting machine that had been donated to the community, Tabacundo, Pichincha Province.

The provision of technical assistance and instruction is a means of counteracting the potential that local people will be unfamiliar with appropriate building design and construction procedures. The reconstruction plans of the national Civil Defense office and the Ecuadorian Housing Bank included the provision of technical assistance and instruction to community residents. Also, books with simple pictures and descriptions on how to repair or to build seismically resistant adobe houses were available for distribution in the communities undertaking reconstruction. Two booklets that were available had been jointly prepared by the Junta Nacional de la Vivienda (National Housing Council) and the United Nations (Junta Nacional de la Vivienda and Centro de Naciones Unidas para Asentamientos Humanos-Habitat, 1987a, 1987b). A third one, on building with *tapia* (sundried mud slabs), had been published after the earthquake by the Centro Andino de Acción Popular (Andean Center for Popular Action) in 1987. It would be useful to document how widely distributed and used such booklets came to be, and how much technical assistance is necessary to support the use of such instructional materials.

Another example of a housing reconstruction program was observed in Imbabura Province (Figure 4.1). There, the League of Red Cross Societies proposed after the disaster to build 1,000 houses to be donated to families selected as the most in need. The project was completed in 1987, with funds permitting the construction of 600 homes. Two designs were used: (1) an urban-style house with a metal frame and block infill and (2) a rural-style house with an exposed wooden beam across the center of the house from which harvested crops could be suspended, as was the custom. For this program, most of the construction materials were purchased locally, and local contractors were used to build the houses, with technical oversight by on-site specialists provided as part of the program (Figure 7.6). Program staff found many families incredulous that someone was going to give them a free house, and one of the initial tasks was to gain credibility with the potential recipients. The varying housing repair and reconstruction programs in the Andean highlands of Ecuador lend themselves to valuable study of the differential social impacts of the various approaches to involving inhabitants in the construction of housing.

In the Sierra, special attention was given to the preservation of various damaged buildings that were felt to have historical and cultural value to the country. They were considered important both for the citizens of Ecuador and for tourism. National funds were used for the repair of these

FIGURE 7.6 League of Red Cross Societies project house, urban style, Imbabura Province.

buildings, most of which were located in the cities of Quito and Ibarra. Governmental funds also were used to repair schools and health clinics, which were the responsibility of national agencies. However, most local public buildings did not come under any of these programs, and there was concern in the smaller communities about how these would be repaired.

Concerns also were voiced about the need for assisting displaced tenants. For example, in Ibarra (Figure 4.3), most of the families left in the tent camps after several months were renters at the time of the earthquakes. As renters, they could not receive assistance to repair their former housing, and usually had to relocate. The housing they had lived in was already the least expensive in the community, and there was very little affordable replacement housing in a market where demand for housing had increased. The tent dwellers had asked to keep their tents, but the local Red Cross chapter had declared its intention to reclaim the tents so they could be used in future emergencies. The camps could not be maintained indefinitely, and it was not clear what provisions were being considered, if any, for supplying housing for these victims.

A loan program to assist homeowners was designed by the Ecuadorian Housing Bank, based on the transfer of some of the bank's funds from regular housing loan programs to programs for earthquake victims. The loans were to be handled by local banks, and the borrowers were able to use the money for housing repair or reconstruction. Initially, it was anticipated that the program could help about 7,000 affected households. However, problems arose over whether or not the bank could use the funds in the manner specified for the loan program, and it appeared that far fewer loans would be possible. Also, as of June 1987, it appeared likely that the initially proposed housing grants for those with the lowest incomes would not be implemented.

The bank estimated that about 70 percent of the affected households had incomes too low to qualify for loans. About 20 percent of the affected households were estimated to have high enough incomes that they qualified only for the normal-rate loans, but many were likely to find the 19 percent interest rate unacceptably high. Only about 10 percent of the victims were estimated to be eligible for lower, preferential, rates established by the program. Thus, the program, as designed, was of limited value.

SUMMARY

At the time of our visit, 3 months after the disaster, the three major problems in the Oriente were: (1) repair of the oil pipeline, (2) deciding what rehabilitation activities were appropriate for the area so clearly identified as hazardous, and (3) deciding on the solution to providing a road that would connect the northeastern part of Napo Province to the rest of the country. Of course, the problem of pipeline repair affected not only the Oriente, but the nation as a whole.

In the Sierra, the time was right for improving the seismic resistance of dwellings in the region. It remains to be seen whether or not the various programs for repairing and replacing damaged housing are implemented widely and effectively enough to significantly reduce the earthquake hazard for the inhabitants of the region.

The observations made in June 1987 revealed many areas in which lessons may be learned from this or future earthquake disasters in developing nations. Recommendations for further research include studies that address:

1. the extent to which technical assistance for construction, both in-person and in the form of written materials, reaches the affected population, and what factors contribute to its effectiveness,

2. the extent to which active efforts are made to distribute special instructional materials on handling the housing and health needs of disaster victims to persons at the local level who can assume responsibility for emergency programs,

3. the development of assistance approaches that emphasize generating local currency versus those that involve providing goods and services by outside organizations,

4. the economic and social consequences of various types of programs for repairing or replacing housing,

5. the relationship between various types of recovery assistance and the promotion of local or national economic development, and policy development for this issue,

6. the effects on reconstruction and relocation decisions of individual victims of providing, or not providing, information about government plans for replacement of infrastructure,

7. policies that help or hurt low-income renters, versus homeowners, whose dwellings were damaged or destroyed, and

8. ways in which the tourist industry can approach the dilemma of not discouraging visitors while not jeopardizing a region's access to assistance, and the extent to which disasters actually affect tourism.

REFERENCES

Bates, F. L. (ed.). 1982. Recovery, Change and Development: A Longitudinal Study of the 1976 Guatemalan Earthquake, Final Report. Athens: University of Georgia.

Centro Andino de Acción Popular. 1987. Construir la casa campesina: hagamos casas de tapia mas seguras. Quito.

Cuny, F. C. 1983. Disasters and Development. New York: Oxford University Press.
Economic Commission for Latin America and the Caribbean (ECLAC). 1987. The Natural Disaster of March 1987 in Ecuador and its Impact on Social and Economic Development. Report #87-4-406, United Nations, Geneva.
El Comercio (Quito). June 17, 1987.
Haas, J. E., R. W. Kates, and M. J. Bowden (eds). 1977. Reconstruction Following Disaster. Cambridge, Massachusetts: The MIT Press.
Hoy (Quito). June 16, 1987.
Junta Nacional de la Vivienda, and Centro de Naciones Unidas para Asentamientos Humanos-Habitat. 1987a. Como hacer nuestra casa de adobe. Quito, United Nations Project ECU-87-004.
Junta Nacional de la Vivienda, and Centro de Naciones Unidas para Asentamientos Humanos-Habitat. 1987b. Como arreglar nuestra casa. Quito, United Nations Project ECU-86-004.

8

Organizational Interaction in Response and Recovery

L. K. Comfort, Graduate School of Public and International Affairs, University of Pittsburgh, Pennsylvania

INTRODUCTION

This chapter examines the organizational interaction that occurred in the response and recovery phases of the Ecuadorian earthquake disaster of March 5, 1987. The organizational interaction is particularly interesting in this event, given the multiple geographic locations of damage from the disaster, the multiple jurisdictional levels involved in disaster response and recovery activities, and the multiple perspectives required for timely and appropriate disaster assistance to the affected populations. Further, the complexity of organizational requirements in some aspects of the disaster operations process increased cumulatively over time, as problems not resolved early in the response phase generated more serious secondary and tertiary effects in the recovery and reconstruction phase.

An organizational perspective offers both value and limitations in a reconnaissance study of disaster operations. Its principal value lies in the presentation of an overview of the disaster operations process, identifying both macro- and microlevel processes in response and recovery. This perspective enables practicing managers and researchers to map the actions that were taken in response to an actual disaster and to examine strengths and weaknesses in these organizational systems in order to improve the design and implementation of future disaster response and recovery programs. The limitations are that organizational interaction in disaster is inherently complex, and this chapter is based upon data that, for reasons of time and resources, may be incomplete.

This chapter presents an initial profile of human systems in action that illustrates the essential integration of administrative jurisdictions, organizational forms, and scientific disciplines in response and recovery operations in disasters. Researchers from other disciplines may find the organizational perspective useful in assessing how information from their respective disciplines affects and extends the capacity of a community to respond to a disaster.

This report serves four objectives:

1. to provide a succinct, descriptive account of the major types of organizations involved in disaster response and recovery activities by geographic location, jurisdictional level, type of authority or funding source, and decision perspective;

2. to present a brief profile of the major organizational networks that evolved into a broadly defined disaster-management system in response to the multiple needs generated by the disaster, indicating the problem focus of the respective networks and the kinds of information needed to support timely, appropriate action;

3. to identify areas of strong performance and areas in need of further development in the set of organizational networks that formed the disaster response and recovery system in operation following the March 5, 1987, earthquakes; and

4. to offer recommendations for further research to improve interorganizational coordination and capacity for action in disaster-management systems.

The Ecuadorian earthquakes of March 5, 1987, produced an unusual set of events that, cumulatively, resulted in a major disaster for the nation.[1] This disaster is an important case for study because it illustrates the interaction between the physical environment, social organizations, economic costs, and human experience. The ensuing set of complex interactions reveals both the opportunities and costs in the development of an integrated disaster-management system, as well as the cumulative effects of problems generated by the disaster across disciplinary fields of expertise and jurisdictional levels.

ASSUMPTIONS

Since disaster operations are very complex, it is necessary to limit this study to aspects that are central to the task of organizing action for disaster response and recovery. This inquiry focuses particularly on the content and exchange of information in facilitating or inhibiting organizational interaction. It is based on a set of assumptions regarding the role of organizations in disaster response and recovery activities, and it is useful to make these assumptions explicit. Briefly, they are:

1. Disasters are complex events, generating multiple physical, technical, social, economic, and psychological needs for the affected populations.

2. Organizations constitute the mechanisms for matching resources and action to needs in disaster response and recovery activities.

3. Effective organizational action depends upon the timeliness, accuracy, and comprehensiveness of information available to decision-makers engaged in disaster response and recovery activities.

4. Information requirements for effective organizational action vary with the decision perspectives and time phases of the disaster-management process.

5. Three functions of the information process—information search, information transfer, and organizational learning—are central to an organization's capacity to take appropriate action in disaster response and recovery activities.[2]

6. Identification of the critical points for information search, information transfer, and organizational learning within and between organizations engaged in disaster response and recovery activities is essential to increasing efficiency and effectiveness in the disaster-management process.[3]

This set of assumptions guided the selection of issues for study as well as the organization and presentation of findings for this chapter. Other issues may be explored in disaster response and recovery operations, but fall outside the scope of this chapter.

ORGANIZATIONAL INTERDEPENDENCE IN THE CONSEQUENCES OF DISASTER

The March 5, 1987, Ecuadorian earthquakes initiated a complex set of events that vividly demonstrate the interdependence of organizational action in response to, and recovery from, the consequences of disaster. Critical problems generated by the earthquakes and associated landslides and floods illustrate important points of choice for informed design of an interorganizational disaster-management process. Further, interaction between the problems triggered subsequent conditions that produced the extraordinary impact of this natural disaster upon Ecuadorian society, as well as the international political economy, through the requirements of technical reconstruction of the Trans-Ecuadorian oil pipeline and economic recovery from the prolonged effects of massive loss of national income.[4] Interdependence in organizational action in disaster response and recovery derives from the complexity of the problems and the sequence of consequences that follow from disruption of major systems such as transportation, economic production, and distribution in the Ecuadorian society.

Domains of Organizational Action

Given the magnitude of the impact of this disaster upon the economic, social, and technical systems of the nation, it is useful to consider the total set of disaster operations as a macrosystem, composed of interacting subsets or microsystems. This step acknowledges the conceptual framework of a nascent disaster-management system that encompasses all response and recovery operations undertaken to assist the affected groups in the Ecuadorian society, while reporting specific projects and programs that were initiated by multiple agencies at various field sites. Within this managerial perspective of the total set of disaster operations, organizational action occurred within three distinct domains or fields of action: (1) geographic location, (2) administrative jurisdiction, and (3) decision perspective. Additionally, organizations as mechanisms of action in disaster operations may be classified by their authorizing structure or mission and funding source into three basic categories: (1) public organizations with legal responsibility for disaster response and recovery, (2) private organizations with responsibility for, and/or interest in, restoring economic activity, and (3) voluntary (private, non-profit) organizations with a mission of humanitarian assistance and social service. This section identifies major problems generated by the disaster by domains of organizational action and illustrates the evolving concept of a disaster-management system driven by interacting events and interdependent processes.

Geographic Location

The earthquakes generated consequences of differing types and magnitudes in three geographic locations of Ecuador, as shown in Figure 8.1.

Zone 1: Western Napo Province

The area of primary impact, which included the epicenter of the earthquake near Reventador Volcano, was located in western Napo Province.[5] In this zone, two major problems were generated by the earthquakes. First, western Napo Province suffered the greatest loss of life, as a result of flash floods generated by massive landslides and debris flows. Official estimates placed the death toll for all three zones of the disaster at 1,000, with 5,000 left homeless or in need of resettlement.[6] Other estimates ranged from 300 to 1,000 dead.[7] While informed sources vary, all experts agree that it is not possible to determine precisely the number of dead, because no reliable census of persons living in the area existed prior to the earthquakes.[8] This observation is particularly relevant for western Napo Province, as it is largely undeveloped territory recently opened to colonists for settlement and cultivation.

Second, major damage occurred to the infrastructure, including destruction of approximately 30 km of the Trans-Ecuadorian pipeline, as well as destruction of approximately 40 km of the main highway from Quito to

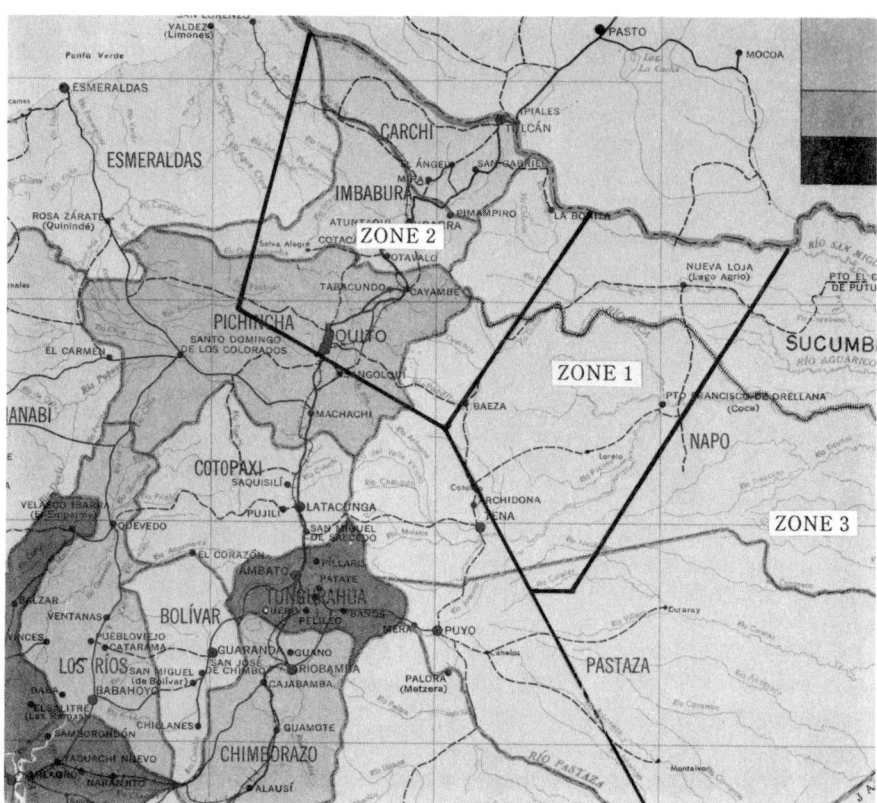

FIGURE 8.1 Three zones of postdisaster operations, March 5, 1987, earthquakes. Zone 1: western Napo Province. Zone 2: Carchi, Imbabura, and Pichincha provinces (highlands). Zone 3: eastern Napo Province.

Lago Agrio, secondary roads, the oil pumping station at Salado, and seven bridges.[9] These conditions are described in detail in earlier chapters and are not repeated here.

These problems precipitated different types of organizational response. At the local level, the first response of the local organizations—the municipal councils, civil defense committees, and parish churches—was directed toward meeting the human needs of shelter, food, and medical care for survivors of the disaster.[10]

After the initial trauma of coping with loss of life and dislocation of families had lessened, a second set of problems centered on the destruction of major infrastructure at the national level. These problems included: (1) reconstruction of the oil pipeline in geologically unstable territory, (2) loss of oil revenues and its consequent impact upon the national economy,

(3) loss of the highway, bridges, and secondary roads for safe travel and economic activity of the resident population, and (4) reorientation and, if needed, resettlement of local residents, severely shaken emotionally and economically by the disaster and struggling to cope with questions regarding an uncertain future in a zone of high seismic risk.[11] The size and scope of this second set of problems necessarily shifted organizational action to national and international levels of operations. The consequences of these operations, in turn, affected the capacity of small communities to sustain their residential populations.

Zone 2: Sierra

The zone of secondary impact from the disaster was the Sierra, specifically the Andean highlands of Imbabura, Carchi, and Pichincha provinces. In this zone, the principal problem was housing. There were no reports of lives lost in the immediate occurrence of the earthquakes, but approximately 60,000 homes were damaged or rendered uninhabitable. The earthquakes produced differential effects for residents of differing economic status. It was an "earthquake for the poor,"[12] as the houses most severely damaged were those made of blocks of sun-dried mud, with no reinforcement or flexibility to withstand seismic movements. Since this is the predominant form of housing construction for the poor residents of the area, the most severe damage from the earthquakes fell on that group within the population already living at a marginal level of economic existence and, therefore, least able to cope with the increased physical, economic, and psychological costs generated by the event. What appeared to be a moderately severe event to other economic groups in this zone proved indeed to be a disaster for those at the lowest economic level, with homes vulnerable to seismic risk and few resources for rebuilding.

A secondary set of problems emerged in these Sierran towns and villages from earthquake damage to the basic infrastructure for community life. Schools, hospitals, churches, and other public buildings suffered structural damage, as did rudimentary infrastructure for water supplies and sewage.[13] Most severely damaged was the mountain city of Ibarra in Imbabura Province, with structural damage to its basilica, schools, and commercial buildings in the town square.[14] Also heavily damaged were municipalities in the cantons of Cayambe and Pedro Moncayo, and neighboring communities in northern Pichincha Province.[15] These types of structural damage adversely affected the capacity of these communities to carry out the normal functions of social and economic life—education, agriculture and commerce, health care, and religious services. Consequently, the burden on individual families to meet daily needs of nutrition, shelter, and physical care was magnified. For those already in marginal circumstances, these needs could not be met by the local systems of economic and social support alone.

Each parish, the administrative unit of local government in these provinces, prepared a summary of the assessed damage from the earthquakes and estimated costs of repair. The parishes, with limited resources, required external assistance to meet these needs and, in turn, referred them to higher jurisdictional levels of government—cantonal, provincial, and national.[16] These requests for resources and assistance for housing and community infrastructure in Zone 2 occurred simultaneously with the demands for assistance from the heavily damaged communities and national infrastructure in Zone 1. Reconstruction needs, experienced first at the local level, escalated to national and international levels of organizational action as managers at successive levels sought economic and technical assistance for rebuilding their communities.

Zone 3: Eastern Napo Province

The third zone of impact from the disaster included the city of Lago Agrio and adjacent communities in eastern Napo Province. These communities suffered little structural damage and had no loss of life. The major problem generated by the earthquakes in this zone was isolation and economic deprivation, resulting from the destruction of the oil pipeline and the major route of land transportation, the highway from Quito to Lago Agrio.[17] These communities survived the initial event of the earthquakes without severe consequences, but the cumulative effects of long-term isolation, unemployment, and lack of access to markets and supplies worsened with the prolonged period required for reconstructing the infrastructure needed to support the local economy, based upon oil production and agriculture.

Especially vulnerable were the Indian communities along the Coca, Aguarico, Dué, Salado, and Papallacta rivers. Dependent upon the rivers for drinking water, nutrition (fish are an important staple in their diet), and transportation, these communities suffered serious deprivation in the loss of vital health and economic resources resulting from the pollution and obstruction of the rivers.[18] The disruption of economic, social, and transportation systems caused by the earthquakes produced, over the succeeding months, a cumulative economic and social disaster for the residents of this zone. It was an interactive set of conditions that, unresolved, steadily worsened and overwhelmed the local resources of the residents and communities of this zone. The local communities could not cope with the conditions generated by the earthquakes without external assistance, and the need for organizational action escalated to provincial, national, and international levels of operation.

Reviewing the consequences of the disaster by geographic location, the earthquakes generated differing types of physical effects in three geographic zones which, in turn, escalated the impact of the disaster for the nation as a whole. Each type of problem required specific kinds of organizational action for appropriate and timely response. The simultaneous needs of the disaster-affected populations in the three zones and the massive impact on the national economy from the combined loss of oil export revenues and costs of rebuilding the pipeline and transportation routes required resources beyond Ecuador's capacity alone.

The interaction of differing needs between[19] the three zones of the disaster, each with its own degree of urgency, compounded the difficulties imposed by scarcity of resources and trained response personnel within this complex organizational environment. The process was dynamic, and in the early phase of disaster response, operating parameters of needs, resources, personnel, and, therefore, action were uncertain. Under these conditions, any nationwide action in disaster management becomes an important measure of capacity. The fact that problems were reported in the coordination and delivery of disaster assistance is not surprising.[20] Rather, the complexity of this situation merits particular attention in understanding the design and dynamics of interorganizational coordination in disaster management.

Administrative Jurisdictions

Jurisdictional involvement in disaster response and recovery activities in the three zones of the disaster reflected the interdependent characteristics of organizational response. Damage from the earthquakes elicited organizational response at five levels of administrative jurisdiction: parochial, cantonal, provincial, national, and international. Although each level served specific functions in disaster response, no single level was able to meet all of the needs generated by the earthquakes alone. Unmet needs at one level pushed the demand for action to the next administrative level, in an escalating search for resources and skills that ranged outside the formal disaster-management system. This section briefly presents the response of the formal disaster-management system as well as the informal response from the private and voluntary organizations committed to humanitarian goals.

In Ecuador, the mission responsibility for disaster management on the national level lies with Civil Defense. This is a developing organization, first established in 1962.[21] The structure for a nationwide civil defense system exists at all jurisdictional levels of administration in Ecuador—parochial, cantonal, provincial, and national. At the local levels, however, civil defense functions are performed largely as an added responsibility for the existing parochial or cantonal councils. The presidents of the municipal councils are, for the most part, also the local directors of Civil Defense. At the cantonal level, the director's position may be held by an officer in the Ecuadorian Army, as in Lago Agrio, where the commander of the Battallon de Selva is also the director of canton's Civil Defense Council.[22] At both parochial and cantonal levels, the national Civil Defense organization seeks links to existing agencies within the communities, for it has few resources of its own. Local governmental units, in particular, assume these responsibilities with little equipment and less training in disaster mitigation and preparedness.

Civil Defense has undergone numerous changes of leadership and direction as it sought to meet the successive challenges of disaster management posed to the nation by the severe earthquake of 1976, the coastal floods of 1982-1983 and the Galapagos fire of 1984.[23] Legal responsibilities are now defined and a clear organizational structure exists, but the capacity of this relatively new organization to take action is limited by the scarcity of resources and trained personnel throughout the intergovernmental system. Although a nation of high seismic risk, Ecuador's disaster management has been limited by the scarcity of resources available for equipment and preparedness training.[24]

The March 5, 1987, disaster, involving three geographic locations with differing requirements for assistance, placed greater demands upon the national Civil Defense organization than its capacity for delivery of services. Recognizing the magnitude of the disaster and its impact upon the nation, President León Febres Cordero Ribadeneira declared a national emergency in the provinces of Carchi, Imbabura, Napo, and Pastaza under the provisions of Article 101 of the Law of National Security.[25] He also called a meeting of the ministers of Health, Finance, Public Works, Energy, Social Welfare, and Environment; the director general of the Ecuadorian State Petroleum Consortium (CEPE); the national director of Civil Defense; the commander-in-chief of the Armed Forces; and the director of the military Corps of Engineers to evaluate the damage and to plan the emergency response.[26] By this action, the president mobilized the highest officers of the Ecuadorian government in response to the disaster, giving it first priority in national affairs. Further, he constituted this group as a national Emergency Committee to direct the disaster operations and to work in conjunction with Civil Defense in the assessment of damage and delivery of disaster assistance. The President appointed Army Gen. German Ruíz, secretary of the National Security Council, as director of the national disaster operations.[27] Gen. Ruíz served as chair of both the national Emergency Committee and the National Center of Emergency Operations. He worked directly with the director of the National Civil Defense Authority, Gen. Antonio Moral Moral. In establishing the national Emergency Committee, President Febres Cordero directly or indirectly engaged virtually all organizations in the Ecuadorian society in response and recovery activities.[28]

Given the limited capacity of the nation and the massive losses generated by this disaster, President Febres Cordero also appealed to the international community for assistance to Ecuador in meeting the technical and economic needs for response and recovery.[29] Some 22 nations responded to this call. The ensuing requirements for coordination and communication between participating nations and between the Ecuadorian levels of governmental jurisdiction in the simultaneous delivery of services to the three disaster zones escalated the complexity of organizational interaction still further.

When the resources and capacity of the respective governmental jurisdictions were unable to meet the urgent needs of the affected communities, responsible citizens turned to private and voluntary sources for assistance. For example, communication is crucial in disaster management, especially in a complex situation involving three geographic areas. Yet, the Civil Defense organizations at the parochial and cantonal levels in Quijos did not have radios to report the news of the disaster and the extent of the damage to the national Civil Defense office in Quito. Telephone communications were disrupted by the disaster, and it took several days for the full extent of the damage to become known at the national level.[30] The only radio available for communication was a station operated by the Evangelical Mission in Quijos, which voluntarily served this important function by relaying messages to and from national offices in Quito.[31]

Voluntary, religious, and communal organizations with appropriate resources and skills responded to humanitarian needs through an informal network of personal and organizational contacts. These agencies sought to provide what assistance they could, but their capacity to do so was limited by lack of resources, equipment, and training for operation in actual disaster environments. Nonetheless, voluntary organizations played an important role in disaster response and recovery, particularly at the community level. In-country organizations joined in a national campaign to offer voluntary contributions to disaster assistance in a remarkable demonstration of solidarity with the disaster victims.[32] Links to the international community provided resources that were not immediately available in Ecuador to initiate the design and implementation of disaster assistance activities. The International Red Cross, Catholic Relief Services, World Vision, and other organizations were important sources of technical expertise, resources, and personnel in the labor-intensive tasks of community assistance to disaster victims. These activities are described in more detail in the next section.

In principle, international and voluntary organizations coordinated their activities with the national Emergency Committee in the conduct of disaster operations. In practice, the complexity of disaster operations and the lack of communications facilities between national offices and provincial and cantonal field sites compelled much of the actual work to be done locally, with limited contributions from the national level.

In summary, public organizations with legal authority for disaster response at parochial, cantonal, provincial, national, and international levels of jurisdiction engaged in disaster response and recovery activities in the three affected zones. The consequences of the disaster, however, overwhelmed the limited capacity of these organizations, leaving the stricken communities with substantial areas of unmet needs. Private and voluntary organizations at each jurisdictional level offered assistance in disaster operations, supplementing and strengthening the capacity of these communities for recovery. International organizations provided resources and skills that enabled local communities to rebuild homes and infrastructure, and the nation to reconstruct the oil pipeline and major transportation routes. The meshing of organizations, across jurisdictions and types of mission, contributed to a developing concept of an integrated disaster-management system.

Decision Perspectives

In order to carry out an effective program of assistance and recovery in any given zone affected by the disaster, response organizations required knowledge and skills from multiple scientific disciplines. The immediate needs of survivors from the stricken communities required medical assistance. The assessment of needs in the damaged communities required knowledge of structural engineering and social welfare. The procurement and distribution of supplies to the affected communities required skills in organization, logistics, and management. The reconstruction of the pipeline, roads, and bridges required technical skills in engineering, geology, and geomorphology. The design and implementation of reconstruction projects required financial, managerial, and political skills and knowledge. Reestablishing households and community services meant matching available resources to immediate needs as appropriately as possible, requiring knowledge of social and economic conditions and organizational skills on a community scale. The jarring instability of the seismic zones left psychological tremors among the population, and a program of recovery to assist residents in overcoming these fears required skills in counseling and psychological assessment. Given the range of problems involved in response and recovery, disaster managers needed expert knowledge and skills from at least five decision perspectives: medicine, engineering, public health, economics, and public policy and management.

In the evolving design of a disaster-management system, the primary organizational task in response and recovery is to connect the elements within each domain of action and to integrate the domains in a coherent, effective program of operations. The objective is to weave a productive network of organizations, drawing resources and skills from relevant jurisdictional levels to meet the needs of the affected communities. This task requires flexibility and resourcefulness at all levels. Information and resources from one domain of action may be helpful—even crucial—to another domain. Rapid, accurate communication between elements of a domain and between domains is critical. This integrative task is demonstrated vividly by the use of *mingas* in community work. The *minga* is a cooperative organization of local residents formed to carry out specific tasks in their communities. It is an organizational form dating from the Incas, and it is well known and accepted in Ecuadorian culture. The needs generated by the disaster, shared by most community residents, created an appropriate opportunity to organize mingas. When provided with limited resources from the national Civil Defense organization, such as a machine for making cement blocks, and supplies from international charitable organizations, such as sacks of cement from the Norwegian Red Cross, residents of communities in the Andean highlands worked in mingas to produce concrete blocks for reconstructing houses for all families in the village.[33] Scarcity of resources and urgency of needs compelled innovative means of organizing action in disaster-affected communities, where responsible managers integrated elements from separate domains to solve a shared problem.

In summary, organizational action in disaster response and recovery activities was directed from the national level by an Emergency Committee established by the President of Ecuador. By creating a separate national committee that had direct access to the experience, expertise, and facilities of the major ministries of the nation, President Febres Cordero in effect distinguished the function of disaster operations from that of disaster planning and coordination performed by the Civil Defense Authority. The two entities worked in close collaboration during the period of disaster operations, but the national Emergency Committee was established as a temporary entity with specific responsibility for the March 5, 1987, earthquakes. The Civil Defense Authority has the continuing legal responsibility for disaster preparedness and prevention activities in the nation. Limited public resources for disaster assistance were supplemented by voluntary contributions of time, expertise, and material from local, national, and international organizations committed to providing humanitarian assistance to the stricken communities. Financial resources for reconstructing the oil and gas pipelines, bridges, and the highway, and relief from external debt obligations exacerbated by the loss of oil revenues were solicited from international sources. Obtaining international monetary credit required interorganizational coordination between nations, a problem that is discussed more fully in the next section.

The striking characteristic of this disaster was the complexity of problems generated by the earthquakes and the interaction among them. The organizational action required to address these problems simultaneously was correspondingly complex. Equally interesting is the degree of innovation with which practical action was developed to use scarce resources at the local level. In the *parroquias* (parishes) of the Sierra and the settlements of Napo Province, the disaster created a policy-making situation, engaging most elements of the community in response to the shared need. For residents living at marginal economic levels, the disaster threatened their basic existence and presented choices for either rebuilding in stronger ways or relocating in geologically more stable territory.

ORGANIZATIONAL NETWORKS IN DISASTER RESPONSE AND RECOVERY OPERATIONS

Reviewing the extent, form, and outcomes of organizational action in disaster response and recovery operations following the Ecuadorian earthquakes of March 5, 1987, a pattern emerges of interacting networks operating simultaneously in reference to particular problems. The degree of communication and coordination within and between these networks varied, with direct effects upon the outcomes of actions taken in the disaster operations process.

In some instances, especially in the rural parishes, communication and coordination between the local, national, and international networks of organizations appeared almost absent.[34] In other instances, as in the emergency housing construction projects undertaken by individual relief organizations in the damaged communities of the Sierra, communication and coordination functioned very well between the local, national, and international levels of operation for a particular project, but failed between projects and between other types of organizations—public, private, or voluntary.[35] In still other cases, communication and coordination between types of organizations and levels of operation developed around problems addressed in common, but failed between sets of problems.[36] In each set of cases, the organizations involved were encountering the limits of time, facilities, and preparedness training essential to achieving effective communication and coordination in the complex, dynamic disaster environment.

The resulting pattern of organizational network performance, at times overlapping, at times operating independently, appears to be a function of at least four factors: (1) overall complexity of the disaster environment, (2) differing requirements of technology and resources for the problems addressed, (3) number and diversity of participating organizations, and (4) limited facilities, staff, and training for communication/coordination in disaster management.

Networks of organizational action centered on the three zones of the disaster, differing in geographic location, physical and climatic conditions, and type of impact upon national, communal, and family life. From the immediate families and communities that suffered physical, personal, and property losses, networks of assistance extended to the parochial and cantonal jurisdictions of public or governmental organizations, but included religious, charitable, and voluntary organizations within the distinct communities. This network of community organizations, in turn, nested within a set of provincial organizations. The provincial organizations served as the linkage between the set of national organizations—public, private, and voluntary—that mobilized resources on the national level and directed them to the problems, communities, and families in the three disaster zones. The network of national organizations responding to the needs generated by the disaster, in turn, functioned within the set of international organizations that contributed financial, technical, material, and organizational assistance to the disaster operations. This overlapping set of networks that characterized the disaster response and recovery operations is represented in the organizational diagram presented in Figure 8.2.

The patterns of organizational action and interaction both overlapped and needed coordination in at least three directions: (1) between local, national, and international levels of jurisdiction and/or disaster operations, (2) between public, private, and voluntary sources of funding and direction at each operational level, and (3) between technical, social, and economic functions served by differing sets of organizations at differing levels of disaster operations. Although breakdowns occurred within categories of organizational operation, performance was reported to be significantly better within a given network than between networks.[37]

Through the national framework for disaster management established by President Febres Cordero to meet the needs generated by the March 5, 1987, earthquakes, the major ministries of the nation were involved in fashioning and implementing appropriate policies for response and recovery, as stated above. Eleven ministries and 10 national institutes or administrative units were included in this national Emergency Committee. Key public officials were also invited to participate in the Emergency Committee, including the President of the National Congress and the President of the Supreme Court. A complete list of the ministries and institutes participating in this committee is included in the Appendixes.

In constituting this Emergency Committee and establishing a National Center of Emergency Operations, President Febres Cordero was, in effect, signaling to the nation that the extraordinary needs generated by the disaster could be met only by a nationwide response, and that the response required the cooperation and contribution of all organizations and citizens. By involving all elements of the nation's political and economic leadership in the directing body for the disaster operations, the President took steps

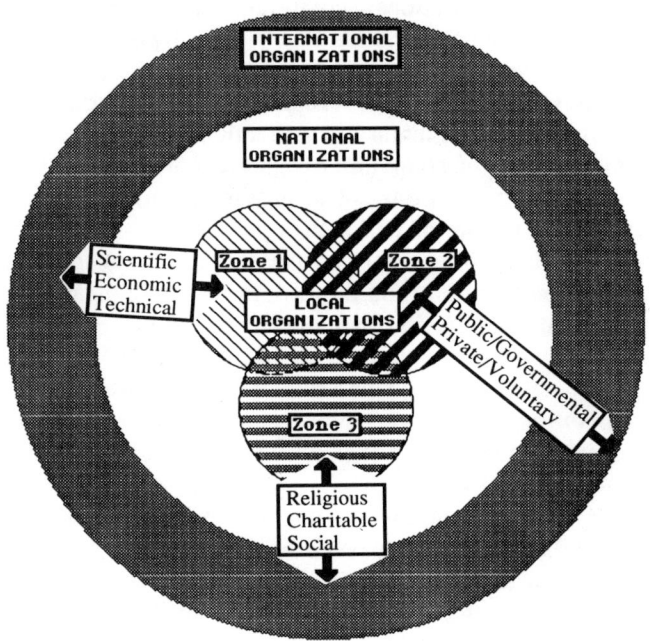

▨ =	Scientific, Technical, Economic Organizations
▰ =	Public/Governmental, Private/Voluntary Organizations
▤ =	Religious, Charitable, Social Organizations
▨▰▤ =	All Organizations

Zone 1: Western Napo Province
Zone 2: Carchi, Imbabura, and Pichincha Provinces, Highlands
Zone 3: Eastern Napo Provinces

While all types of organizations played some role in each disaster zone, the dominant pattern of organizational interaction is indicated for the respective zones. The overlap in patterns represents the coordinative processes between jurisdictional levels.

FIGURE 8.2 Networks of organizational action in postdisaster operations: the March 5, 1987, Ecuadorian earthquakes.

to provide credibility to the concept of a nationwide network of emergency response.

The effectiveness of this network of emergency operations in serving the needs for organizational coordination in disaster management is a question for further study. There is evidence of both success and failure of this system, assembled under stress of disaster, to achieve coordination in response and recovery activities. This report serves only to state the problem and to offer an initial account of its operation in the overall disaster-management process. Clearer insight into the problem of interorganizational coordination in disaster response and recovery activities can be illustrated by a brief profile of organizational activities undertaken to meet the major problems identified in the three disaster zones.

Disaster Operations Network 1: Western Napo Province

In western Napo Province, the zone of primary impact, three major problems generated by the disaster required urgent, simultaneous response and assistance. Moreover, the three problems, each separately traumatic, interacted to rapidly escalate the demands for organizational response.

The most urgent need was for food, clothing, shelter, and medical care for the victims of the disaster, those families who had suffered physical damage and/or lost their homes and property in the chain of destructive events generated by the earthquakes. Given the undeveloped state of the territory, the lack of communications facilities, and the marginal economic conditions of the resident population, the task of meeting basic human needs for the victims required outside assistance. This task, however, was made more difficult by the rudimentary state of public services within the communities and the lack of equipment, training, or experience with disaster management. In most communities, the local churches had established stronger relationships with the citizens than had the Civil Defense councils, which were relatively new and still in the process of development. Individual members of community Civil Defense councils were willing to participate in disaster-response activities, but the responsibilities had not been clearly defined at the local level, and there were few resources or experienced personnel to guide the process.[38] Consequently, citizens looked to trusted religious leaders for practical as well as spiritual guidance.

Bringing assistance into the local communities from national and international sources generated all of the problems of unplanned action in complex, dynamic environments. Local managers reported a series of dilemmas as they sought to assess needs and distribute relief goods appropriately in their respective communities. National and international organizations contributed supplies, but how was this assistance to be distributed and to whom? Was it ethical to make distinctions between levels of economic misery, as some families who had not lost their homes in the disaster suffered just as acutely from the ensuing loss of employment or income from agricultural products? How did one serve the intangible needs of the people in the wider community for emotional security in a geologically unstable environment? The immediate needs for physical care and basic family support for the disaster victims were compounded by the economic needs of all members of the community, as well as increased difficulty in communication and transportation.[39] These questions involved the national

Civil Defense Authority; the ministries of Health, Social Welfare, and Agriculture and Livestock; the Ecuadorian Institute of Children and Families; and the Ecuadorian Institute of Agrarian Reform and Colonization; as well as the Ecuadorian Red Cross; Catholic Relief Services; Hoy Cristo Jésus Benedictus (the service organization of the Evangelical Church); U.S. Peace Corps volunteers; and other international organizations that contributed disaster assistance.[40]

Basic needs for the disaster victims in western Napo Province were overshadowed by the stunning blow to the national economy from the destruction of the Trans-Ecuadorian oil pipeline and the consequent lapse in oil exports. The second problem, reconstruction of the pipeline and highway in a geologically unstable area, required involvement of a different set of organizations—national and international—to address the scientific questions of probability of seismic risk, the landslide and flood hazards, engineering questions of feasibility and design, and economic questions of costs and credit for the project. This problem drew the attention, involvement, and cooperation of a number of national and international organizations to determine the financial and technical feasibilities of reconstructing the pipeline and highway, the implications for national economic performance of alternative routes, and the costs associated with each alternative. These organizations included those with scientific expertise, such as INEMIN (Ecuadorian Institute of Mining); the Institute of Geophysics of the National Polytechnic School; the Italian Mission, with its team of geophysics experts; and the U.S. Geological Survey; as well as those with access to financial credit, such as the World Bank, the Interamerican Development Bank, the Andean Development Corporation (CAF), and the International Monetary Fund.

The problem also involved substantive negotiations between national representatives, such as the President of Ecuador and the Vice-President of the United States, in securing financial credit. Further, it involved negotiations between the presidents and appropriate ministers of Ecuador and Venezuela, as well as representatives of the Organization of Petroleum Exporting Countries, to arrange Venezuela's assumption of Ecuador's oil production quota and export obligations, between Ecuador and Colombia to arrange Ecuador's use of the Colombian pipeline to maintain partial production and shipment of oil during the reconstruction period, and between Ecuador and Mexico to arrange for technical and material assistance in the reconstruction of the pipeline.[41]

Negotiations also occurred between particular organizations, such as between CEPE (Ecuadorian State Petroleum Corporation) and Will Bros., the U.S. engineering firm that built the original pipeline; and CEPE and Texaco Oil Co., the U.S. company involved with the Ecuadorian national corporation in the production and shipment of oil, as well as the maintenance of the related facilities. This network of scientific, financial, business, and governmental organizations working on the questions of reconstruction of the pipeline in order to resume oil production and exports operated largely independently of the set of organizations involved in humanitarian disaster assistance. Yet, the consequences of their actions affected the residents of the communities through the loss or generation of jobs, access to transportation, and the benefits of remaining in the region.

A third major problem generated by the disaster in western Napo Province concerned policy decisions for future economic and agricultural activity in the zone. The question of reconstructing the infrastructure—pipeline, roads, and bridges—was tied to scientific information regarding the geological stability and probability of future seismic and/or volcanic activity in the zone. The feasibility of resettling the colonists living in the zone into areas of lower seismic risk involved judgments of economic costs versus social responsibility. The cost of the ecological damage to the rivers and the short- and long-term effects of this damage upon human populations in the region also required multidisciplinary study and design for action. These questions, dealing with the viability of continuing economic and social activity in the region, were the most difficult and most problematic in terms of policy design and implementation. Decisions made regarding infrastructure in the early stages of disaster response and recovery shaped the formulation of future options and policies for development of the region.

Organizations involved in the process of determining the future development of the zone, and alternatives for its current residents, were the scientific organizations—INEMIN; Institute of Geophysics of the National Polytechnic School; the Italian Mission, with its team of experts in geophysics and volcanology; the U.S. Geological Survey; and the ministries of Agriculture and Livestock, Social Welfare, Health, Environment, and Transportation; IERAC (Institute of Agrarian Reform and Colonization); the Ecuadorian Institute of Children and Families; as well as the private and voluntary organizations—Catholic Relief Services/CATEC (Corporation of Support to Technology and Communication), environmental groups, and the provincial and national Indian associations (CONFENAIE or Confederation of Napo Associations of Ecuadorian Indians and CONAIE or National Confederation of Ecuadorian Indians).

In summary, three problems generated by the disaster in western Napo Province interacted with one another to produce an even more complex and difficult set of policy questions for organizations engaged in disaster response and recovery. These were the needs for: (1) immediate humanitarian assistance to the victims and their families, (2) reconstruction of the infrastructure required to resume production and export of oil for the national economy, and (3) review and redesign of future economic and social activities for the zone, taking into account continued seismic risk. Sorting out constructive alternatives for appropriate action was clearly an interorganizational task. No single agency or entity could manage it alone, and the evolving pattern of actions included local, national, and international agencies; public, private, and voluntary efforts; and scientific, technical, administrative, economic, and social concerns. The complexity of tasks in disaster response and recovery greatly increased the difficulties of designing coordinated organizational action.

Disaster Operations Network 2:
The Sierra—Pichincha, Imbabura, and Carchi Provinces

The major problem generated by the earthquakes in the Andean highlands of Pichincha, Imbabura, and Carchi provinces was the loss of housing, exacerbated by the prevalence of poverty. In Canton Cayambe of Pichincha Province alone, nearly 3,000 houses were damaged or destroyed, leaving 15,000 people with minimal shelter in cold, rainy weather.[42] In all three Andean provinces, approximately 60,000 homes were reported destroyed or damaged, or 81.8 percent of the total number of 73,261 houses affected in all zones of the disaster.[43] Damage ranged from total destruction to small cracks. In this zone, approximately 9,566 homes needed to be wholly reconstructed, the largest proportion of the total number of houses, 11,694, reported in need of complete reconstruction for the entire disaster. Approximately 47,828 people, residents of the damaged houses, were left homeless in this zone, out of the 58,470 persons reported homeless for the total disaster. In addition, some 33,947 persons in this zone suffered the cost and inconvenience of repairing damage to their homes, out of the 41,500 persons reporting repairs for the entire disaster.

In sum, over 80 percent of the damage to housing generated by the disaster was reported in the Andean highlands, affecting some 81,755 people, out of the approximately 100,000 people who reported damage to their homes for all zones of the disaster.[44] The extensive need for housing in this zone generated organizational response at the local, national, and international levels and involved public, private, and voluntary organizations. Technical, economic, and cultural perspectives influenced the design of specific programs of action to meet housing needs of the disaster victims, as well as the acceptance and implementation of the various programs.

Local organizations initiated the response to housing needs for community residents who either lost their homes or suffered serious damage to their homes in the earthquakes. While the extent and efficiency of the response varied by community, local councils in most communities undertook the census of homes, identifying those families who had suffered damage in the earthquakes.[45] Many communities, however, did not have the resources to rebuild or repair the damaged homes. Consequently, they turned to provincial and national sources for assistance, which in turn requested assistance from international sources.[46]

International organizations played a major role in the reconstruction of housing in this zone. First, housing was an obvious and tangible need that they could meet by providing materials for shelter and for design and reconstruction for the residents of these rural communities. While there was an initial effort to communicate with one another and to coordinate their assistance programs, the various international organizations proceeded to carry out their work, for the most part, independently. The USAID/Office of Foreign Disaster Assistance distributed plastic sheeting widely in the area as an immediate and temporary protection to families sleeping outside.[47] The British government sent tents. The Norwegian Red Cross sent construction materials. The International Committee of the Red Cross, working in conjunction with the Ecuadorian Red Cross, designed and constructed a sizable housing project near Ibarra. The German and Italian governments selected certain communities and met housing needs in those communities. Other governments and other international voluntary organizations contributed construction materials, technical assistance in architectural design, and money to assist in the rebuilding or repair of housing in these mountain communities. In Cayambe, for example, all housing construction and repair was committed for construction and financed by international or national sources by June 30, 1987.[48]

Disaster assistance from foreign governments was sent to the Ecuadorian government and transmitted to COEN, the National Center for Emergency Operations. The Civil Defense organization, working with COEN, was responsible for distribution of international assistance from foreign governments to needy communities in the disaster zones. Assistance from international voluntary organizations such as World Vision, Friends of America, and Save the Children were transmitted to a committee established by the national government specifically to receive international disaster assistance. This committee was chaired by the First Lady of Ecuador, Sra. Eugenia Febres Cordero. The International League of Red Cross Societies and foreign national Red Cross societies worked directly with the Ecuadorian Red Cross to conduct its extensive housing program. Other international organizations, such as Catholic Relief Services and the Evangelical Brotherhood, transmitted money, technical assistance, and materials for housing through their respective in-country organizations.

At the community level of disaster operations, the local churches were frequently the primary vehicle for distribution of supplies or mobilization of services to assist the disaster victims. Long established in the communities and trusted by community residents, religious leaders from local orders served a very important function in linking the needs of the residents to the sources of assistance from the international community.[49] In the process, they served the equally vital function of providing reassurance and hope for rebuilding their lives to community residents, badly shaken emotionally by the sudden experience of destruction in their lives.[50]

In summary, organizational interaction was important in addressing the major problem of housing in the disaster zone of secondary impact. Although international organizations played a major role in the reconstruction projects initiated to meet this need, they operated largely independently, with little communication and coordination between projects. Questions of design, cost, and appropriateness of housing for the local needs of community residents, involvement of residents in the actual development of the projects, and work of reconstructing their own homes were recognized by local leaders but were not addressed systematically among the set of participating organizations.[51]

The difficulty of establishing a viable means of coordination during the actual operations was also recognized by participants in the process who had tried to do so.[52] The opportunity to use the reconstruction process to serve other community needs was very apparent in the housing projects of the Andean highlands. Important initiatives were taken, but other problems surfaced in the process. The appropriate role of communities in designing and building their own housing needs to be carefully examined. The linkage between the destructive consequences of disaster and constructive opportunities for fostering community development in the reconstruction phase are clearly illustrated in the housing projects undertaken in this zone of secondary impact.

Disaster Operations Network 3: Eastern Napo Province

In eastern Napo Province, the major problems generated by the disaster were the interaction of isolation and unemployment resulting from the destruction of infrastructure—the highway and the oil and gas pipelines—and the pollution of the rivers, with its consequent impact upon health and agriculture. These problems were especially difficult because they deepened with time and were compounded by decisions made in reference to other aspects of disaster recovery and reconstruction. Destruction of the oil and gas pipelines meant unemployment for many of the residents of these eastern communities, dependent directly and indirectly upon petroleum production as the primary local industry. Further, decisions made in reference to long-term development for Zone 1 and the reconstruction of the highway and bridges in geologically unstable territory adversely affected access to markets and supplies for the population in the eastern provincial cities and settlements. The primary means of transportation between the eastern cities of Lago Agrio and Coca and the metropolitan markets of Quito were by air, which was expensive, or by water through a network of rivers that led to roads, which was time-consuming.

Networks of organizational assistance did develop to address these problems. However, these networks appeared to cluster even more specifically around the problems, with little coordination between them or recognition of their mutual contributions to the overall disaster response and recovery effort. First and most widely recognized was the *Punta Aérea*, the aerial bridge operated by the Ecuadorian Air Force between Quito and Lago Agrio with international assistance. The governments of Great Britain, Argentina, the United States, and other nations contributed time of aircraft and crews to assist the Ecuadorian Air Force in the transport of supplies, commodities, and people to and from the isolated cities.[53] A major factor inhibiting the use of air transportation, however, was the cost, calculated at $600-800 per hour for a military helicopter with limited load capacity and many times that for a C-130 or similar cargo aircraft.[54] Consequently, foreign governments limited their contributions to specific periods of time or reduced their participation as the months passed. Without viable road transportation, however, the isolation worsened the economic and social conditions for the populations in these eastern communities, which were marginal even before the disaster.

A second network developed among the Indian organizations and religious organizations that offered services to the Indian communities and colonists who lived in the rural areas or along the rivers of eastern Napo Province. These populations, largely unconnected with the national economy and structure of public service organizations, had already developed a network of relationships focusing on self-help and voluntary assistance. These organizations included: (1) the Federación de Organizaciones Indígenas de Napo (FOIN), which has a strong membership base near Puyo; (2) the Confederación de Nacionalidades Indígenas del Ecuador (CONAIE), the national confederation of Indian organizations; and (3) the Federación de Comunas de Napo Ecuatoriana (FECUNAE), which has strong influence in the settlements along the Coca River.[55] The ecological effects of the disaster in the pollution and obstruction of the rivers had particularly severe consequences for these populations, dependent upon the rivers for food, water, and transportation. Again, living in largely undeveloped territory on a marginal economic base, these populations were particularly vulnerable to the chain of adverse consequences generated by the earthquakes. Their limited resources to cope with disaster conditions were quickly exhausted, and they needed outside assistance for recovery and reconstruction.

The primary link between the Indian and communal organizations of eastern Napo Province and the national and international sources of disaster assistance were religious and voluntary organizations. For example, Misión Carmelita, located outside Lago Agrio along the Coca River, took an active role in contacting the Indian communities along the rivers and in organizing the distribution of supplies and assistance to them.[56] Catholic Relief

Services also utilized contacts already established through a program of technical assistance to improve economic and social conditions in these undeveloped communities.[57] With long experience in the Indian communities and language capacity in Quecha, the Catholic religious leaders were able to engage in a needs assessment following the disaster and the design of strategies for self-help with the residents of these isolated communities. An Evangelical mission near Lago Agrio played a similar role. These mission leaders were able to articulate the needs generated by the disaster in their respective communities to the national and international organizations in order to mobilize needed supplies and assistance for recovery and reconstruction activities. The religious and voluntary organizations thus served a vital role in linking the local needs with national and international sources of assistance in disaster operations in eastern Napo Province.

A third network evolved to address the problem of isolation for eastern Napo Province between the Ecuadorian Army Corps of Engineers, the Ecuadorian Ministry of Public Works, the USAID/Office of Foreign Disaster Assistance, and the U.S. Military Group in Ecuador, which included U.S. military personnel from the Army (including the Corps of Engineers, and the Southern Command based in Panama) and the Air Force. This network formed around two road construction projects to open up new southern routes from Quito to the settlements and cities of eastern Napo Province. In the first project, the Office of Foreign Disaster Assistance of the U.S. Agency for International Development agreed to purchase the materials and to task the U.S. Army Corps of Engineers to construct 11 bridges needed for a road being built by the Ecuadorian Army Corps of Engineers for the Ministry of Public Works (Ministerio de Obras Públicas, MOP). The road, already under construction when the disaster occurred, was planned as an alternative route from Quito to Coca through territory that was geologically more stable and less vulnerable to seismic risk than the Quito-Lago Agrio highway. The planned route, however, traversed rugged mountain and jungle terrain, and the estimated completion date for the road, given existing Ecuadorian resources, was 3 years away. The completion of the road would be shortened by at least 2 years with the provision of the seven bridges needed to cross the rivers in the area. As a major contribution to Ecuadorian disaster recovery and reconstruction efforts, the Office of Foreign Disaster Assistance agreed to finance the construction of the needed bridges. The construction of bridges, however, was diverted from the MOP road built by the Ecuadorians to a second road project, still farther south, named the "Blazing Trails" project, which was to be built as a training project for U.S. Army reservists.

The controversial Blazing Trails project, conducted by the U.S. Army in cooperation with the Ecuadorian Army Corps of Engineers, was a second effort in Ecuadorian-U.S. intergovernmental cooperation in disaster assistance. The project entailed building a second road connecting the existing road system from Quito to the eastern cities of Coca and Lago Agrio in Napo Province, located farther south from the MOP road. The project was designed as a U.S. military exercise in jungle training conducted under contract and with the legitimate approval of the Ecuadorian government. It would utilize the services of U.S. Army reservists to meet a civil engineering need for Ecuador. On the basis of the planned operation and the information available to them, OFDA decided to withdraw the bridges from the MOP road and install them where needed for this second road. This decision engendered substantial political controversy in the Ecuadorian National Congress.[58] Misperceptions and distrust of U.S. military purposes in bringing reservists into the country for training created an uneasy tension in Ecuadorian circles over the project. The difficulty of the construction conditions and the inexperience of the reservists working in jungle terrain caused policymakers at OFDA to reconsider their decision on the location of the bridges in terms of where they would contribute the most to reopening transportation in the area.

After reviewing the Blazing Trails road project and construction conditions more carefully, OFDA moved the bridges back to the MOP road to the north.[59] Given its goal of facilitating transportation to the isolated region of eastern Napo Province, OFDA policymakers concluded, for several reasons, that the MOP road would likely be completed more quickly than the southern route. The political context of the Ecuadorian presidential elections of January 1988, anticipated in the developing campaign strategies, may have influenced the debate,[60] but the controversy distracted time, energy, and attention that might have been put to more substantive cooperation between U.S. and Ecuadorian organizations in disaster assistance.

The project is interesting because it illustrates a critical dilemma regarding the role of information in building the basis of common understanding necessary for interorganizational coordination in the dynamic context of international disaster assistance. The controversy illustrated the dilemma of technical versus cultural exchange of information between the two governments regarding the construction of the road, and the ensuing constraints upon organizational interaction. To inform the Ecuadorian public regarding the cultural/humanitarian objectives and means of the U.S. military road construction project in reference to the overall goal of disaster assistance took scarce time and attention from technical work on the project. Yet, to do technical work on the project without fully informing the Ecuadorian people regarding its cultural/humanitarian intent raised questions of public trust, which inhibited the overall goal of disaster assistance. Within the complex, uncertain environment of disaster operations, the dilemma illustrates the importance of information search, information transfer, and organizational learning in creating a common understanding of the goal of disaster assistance in international projects sensitive to the possibility of dependency.

This agreement demonstrated an important step in productive interorganizational coordination between Ecuador and the United States on a critical technical problem. Further, it demonstrated important interorganizational coordination within the U.S. Mission in Ecuador to utilize the technical capacity of the Army Corps of Engineers (Department of Defense) to meet the humanitarian goals of international disaster assistance (Department of State).

In summary, three distinct networks of organizational interaction developed in relation to the problems of isolation and transportation created by the earthquakes in eastern Napo Province. These included: (1) the network of national governments that contributed support to the Ecuadorian Air Force in its maintenance of the aerial bridge between Quito and Lago Agrio, (2) the network of religious and voluntary organizations that linked the communities of Indians and colonists to national and international sources of disaster assistance, and (3) the network of Ecuadorian and U.S. governmental organizations involved in technical assistance for road construction through remote areas of eastern Napo Province. Critical, however, from the standpoint of achieving interorganizational coordination was the relatively low degree of communication and coordination among the three networks, even though each developed in response to the common problem of isolation and transportation for inhabitants of eastern Napo Province.

Returning to the concept of an emerging disaster-management system, linkages developed through both formal and informal contacts among the three disaster operations networks. The set of networks evolved from the practical requirements of action in disaster response and recovery. Taken together, they offer an initial basis in experience and adaptive skills that may inform and reinforce an integrated disaster-management system for Ecuador. The set of networks is represented in Figure 8.3.

Strengths and Needs for Further Development in the Larger Network of Disaster Response and Recovery Operations

Strengths

Many strengths emerged from the set of organizational interactions that characterized the disaster response and recovery operations following the March 5, 1987, earthquakes. Five practices, in particular, deserve mention, as they either confirm basic principles of disaster management or indicate innovative practices developed in this disaster:

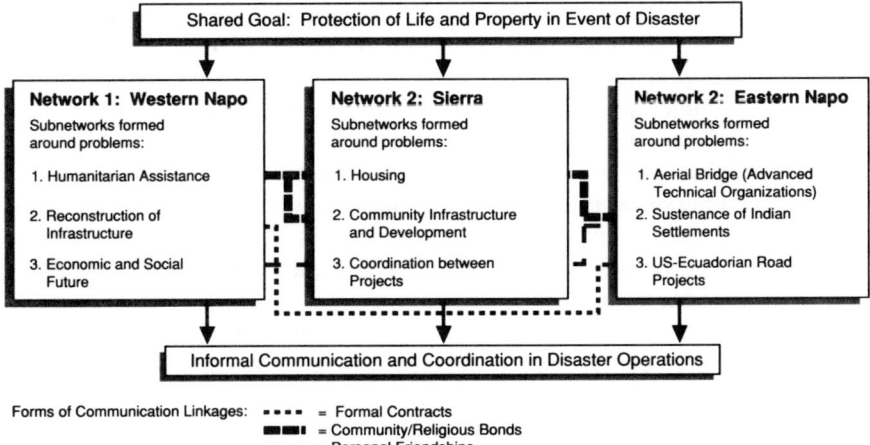

FIGURE 8.3 Evolving disaster-management system: March 5, 1987, Ecuadorian earthquakes

1. This disaster drew an extraordinary response in voluntary contributions from in-country organizations. An outpouring of goodwill is not unusual following a major disaster, but this response deserves mention because it was carefully mobilized and directed through the use of a nationwide television campaign with international transmission. Virtually all major organizations in Ecuador—public, private, and voluntary—participated in the "Crusade for Solidarity" to raise money and contributions for disaster assistance. The appeal, carefully structured, focused not on sympathy for the victims but on the national sense of unity in meeting the unexpected hardships of disaster. Measured by commitment to action as well as contributions, the campaign was very successful.

2. Review of disaster operations demonstrates again the fundamental value of respecting local institutions, practice, and knowledge in engaging local residents in the difficult tasks of rebuilding their lives. The productive work of the *mingas* in constructing new homes, the value of family and friendship ties in overcoming the despair of sudden loss, and the knowledge of local officials and field directors of development organizations in designing effective reconstruction strategies all indicate the importance of fitting the requirements of disaster management to local delivery systems.

3. The importance of prior networks of common goals, shared work experience, and professional association in facilitating the mobilization of resources and action across jurisdictional lines was vividly demonstrated by the ease with which communication traveled and action flowed *within* such networks as Catholic Relief Services, USAID/OFDA, Red Cross

societies, Ecuadorian and international military experience, and family and community ties. Each generates the critical factor of trust, so essential to decision making in the uncertain conditions of disaster. Conversely, the absence of common experience is illustrated in the frequent breakdown of communication and coordination *between* these networks, resulting in frustrating difficulties in the implementation of disaster-assistance programs.

4. Needs assessment is a time-honored technique, but how to do it promptly, accurately, and effectively to serve as a basis for urgent action constitutes a persistent problem in disaster management. The strategy of having cash grants available to place an interdisciplinary team of experts immediately in the field worked very well for Catholic Relief Services. Prior experience in the country, prior experience in disaster, and pooled knowledge from multiple disciplinary perspectives resulted in a solid information base that enabled CRS/CATEC to design a field program of emergency assistance and rehabilitation that proved very effective in Napo Province. The model merits replication.

5. A step-by-step approach of engaging the victims of disaster in the difficult process of rebuilding their own lives and communities was demonstrated repeatedly to be more effective than inviting dependency through continued distribution of aid without acknowledgment or return investment. The trauma of disaster is unsettling at best, and to people with marginal resources and uncertain futures, it can be devastating. A combination of care, technical assistance, and clear guidance through incremental steps worked very well in the community programs instituted in Cayambe through the community's Emergency Committee and the CRS/CATEC program in Napo Province. Conversely, the poorly coordinated distribution of relief supplies by external organizations often had adverse effects in disaster-stricken communities, resulting in negative competition for goods and greedy distortion of personal needs.

Needs

Clear needs for further development in disaster management were also demonstrated in this set of disaster operations. They include:

Improved Communication

Improving communication is the most critical need for increasing effectiveness in disaster management. Organizations cannot function well without knowing the needs, resources, limitations, and time sequence for action, both within their own spheres of responsibility and between the multiple organizations engaged in the process. This process can be accomplished only through open, interactive communication among the responsible leaders engaged in disaster response and recovery activities. Further, the process works best when it has been designed and practiced *prior* to the disaster event. Facilities, equipment, and training are needed to improve communication, especially at the local level of operations, where the burden of responsibility is highest but resources are most limited, and between jurisdictional levels of operation.

Increased Coordination of Action

Increasing coordination between the multiple organizations engaged in disaster operations is critical to improving performance. The functions of communication and coordination are complementary, and both are consistently vulnerable to failure in uncertain, complex disaster environments. A shared knowledge base of information, resources, skills, and participants in the disaster-management system is critical to enable specific organizations to adjust their performance, respectively, to others engaged in the process of coordinated action toward the common goal of rescue of human life and restoration of property.

Advanced Information Functions

Information functions for the wider set of organizations involved in disaster management need to be developed more fully and more systematically. In this disaster, many organizations conducted separate information searches, but there was difficulty in sharing information and especially in obtaining timely, accurate information from rural disaster sites. Representation of information in a common format and transmission of information to multiple participants in a timely manner also are critical functions in improving the communication and coordination processes central to effective performance in disaster management.

Revised Concept of Disaster Assistance

The allocation of assistance to victims of disaster is especially sensitive in communities with marginal economic standards. The design and distribution of assistance need to be reconsidered in terms of the opportunities it creates to rebuild lives and homes in stronger, more productive ways. Organizations that incorporate planning for a stronger future with immediate relief from the destruction of disaster are effective in mobilizing not only individual families but wider participation in community programs to reduce vulnerability to disaster.

Organizational Learning in Disaster Management

Evaluation of performance in disaster operations and constructive feedback to participating organizations is essential in developing the capacity of these organizations to mitigate conditions that may contribute to future disasters. Developing a communitywide orientation toward reduction of risk and knowledge of emergency procedures is a vital step toward fostering responsible civic action in a zone of high risk from earthquakes and other natural hazards such as landslides and floods. This feedback is essential for international organizations as well, for they play a critical role in disaster operations in developing countries. Because organizing effective participation at the international level is necessarily more complex, feedback in a careful, constructive format is especially valuable. Without design, organizations are unlikely to learn from past experience, and may repeat ineffective performance in future disasters.

Recommendations for Further Research

The fundamental problem confronting organizations in disaster operations is to design effective action in this uncertain, complex environment. Four recommendations for future research appear especially promising in developing organizational capacity to improve performance in this difficult, dynamic environment. Each focuses on information content and exchange as the most productive and least expensive means of improving capacity for organizational learning and interorganizational coordination in disaster conditions. They are:

1. The design and development of an interactive information system for decision support in disaster management. Such an information system would be based on an interdisciplinary knowledge base with provision for interactive communication among multiple users. This research would utilize new developments in information, radio, and telecommunications technology to address recurring problems of communication and coordination.

2. The design of interorganizational and interjurisdictional simulated-disaster-operations exercises as a means to explore the limits and capacities of human decisionmaking processes in disaster environments. Such exercises might utilize an interactive information system to explore the linkage between information technology and organizational learning in complex environments.

3. Inquiry into the design and development of networks as appropriate organizational forms for the rapid mobilization, implementation, and evaluation of action in disaster management. Such networks might cross disciplinary, organizational, and jurisdictional lines and would be designed to facilitate action in this complex environment. The basis for each network

would be shared knowledge of particular problems in disaster operations or phases of disaster management. A network of networks might organize this information in a model of disaster management which, in turn, could be incorporated into a computerized, interactive information system.

4. Inquiry into economies of resource management that would facilitate interorganizational participation. Resources might be allocated to solve specific problems, such as transportation, and organizations with relevant skills, equipment, and capacity to address this problem could draw against the account established for designated participants in that network. A computerized information system would facilitate the monitoring of expenditures for a multiorganizational project.

ACKNOWLEDGMENTS

The research for this chapter was supported by funding from the National Research Council, as well as the Center for Latin American Studies and the Office of Research of the University of Pittsburgh. The author gratefully acknowledges these organizations for their financial support. In addition, many people in both Ecuador and the United States contributed time, effort, and assistance to the conduct of this study, and I thank them all.

A number of people in particular offered valuable direction, insight, and guidance to this study, and I am deeply grateful to them for their assistance. They include: Dr. Blasco Penaherrera, Vice Presidente de Ecuador; Ing. Horacio Rueda, Director General, INEMIN; Gen. Antonio Moral Moral, Director, Defensa Civil de Ecuador; Ing. Hernan Orellana, Ing. Renan Herrera, and Ing. Raul Montalvo, of INEMIN; Prof. Guido Zambrone, Ministry of Finance and Universidad Católica; Prof. Alvaro Saenz, Facultad de Latin América de Ciencias Sociales and INFOC; Maj. Luís Aguas, Comandante de Battallon de Selva; Mario Venegas, CATEC; all of Ecuador. At the U.S. Mission in Quito, Neil Meriwether, Ricardo Bermudez, Gordon Jones, Col. Paul Scharf, Col. Troy Scott, Maj. Howard Mayhew, and Capt. Robert Parsons were especially helpful. At the University of Pittsburgh, Amy Jacob, Lynn Whitlock, Keun Namkoong, and Elizabeth Bermant helped me to manage the daily tasks involved in the conduct and analysis of the research. While all have generously offered guidance and assistance, any errors in fact or interpretation are those of the author alone.

NOTES

1. Hoy, Quito, Ecuador, March 10, 1987, p. 1. President León Febres Cordero Ribadeneira stated that " . . . this is the most serious disaster in the history of Ecuador as a nation."

2. These functions are nontechnical terms that describe practical activities undertaken by any manager in confronting a novel set of events that requires organiza-

tional action: searching for the best information available to assess the situation before taking action; communicating that information to the relevant persons or organizations involved in, or affected by, the proposed action; and evaluating the effects of the actions taken in order to determine the next appropriate steps. For a fuller discussion of these terms, see Chris Argryis. 1982. Reasoning, Learning and Action. San Francisco: Jossey-Bass.

3. These assumptions are drawn from previous research in disaster management, problem solving, and organizational theory. They rely on the work of many authors, but especially Herbert A. Simon, The Sciences of the Artificial (Cambridge: The MIT Press, 1969, 1981); Allen Newell and Herbert A. Simon, Human Problem Solving (Englewood Cliffs, N.J.: Prentice-Hall, 1972); Russell Dynes, Organized Behavior in Disaster (Columbus, Ohio: Heath-Lexington, 1974); Harold Linstone, ed., Multiple Perspectives for Decision Making: Bridging the Gap between Analysis and Action (New York: Elsevier, 1984); Anthony Debons, as presented by Isabel Cilliers, Problems in information science, Information Age, Vol. 7, No. 3, (July 1985):150-155; John Holland, Adaptation in Natural and Artificial Systems (Ann Arbor: University of Michigan Press, 1975); and Louise K. Comfort, The San Salvador Earthquake, in Uriel Rosenthal, Michael T. Charles, and Paul t'Hart, eds., Beyond Crises (Chicago: Charles C. Thomas, 1989).

4. Hoy. Accounts of ecological, technical, economic, social, political, cultural, and international impacts of the earthquakes were reported daily in the two major Quito newspapers, Hoy and El Comercio, during March 1987 and continuing during the succeeding months. The author read both papers daily during the period of her field study, June 14 – July 15, 1987, and sought to obtain back issues of both papers for the month of March 1987. Regrettably, she was unable to obtain a complete set of back issues of El Comercio for this period. Consequently, the newspaper references in this analysis are drawn primarily from Hoy during March 1987, but refer to both newspapers during June and July 1987. To counter any possible bias from a single source, the author sought to find at least two references for critical points in the analysis.

5. See maps of disaster zones presented in earlier chapters.

6. United Nations Economic Commission for Latin America and the Caribbean (ECLAC). The Natural Disaster of March 1987 in Ecuador and its Impact on Social and Economic Development. Report #87-4-406, May 6, 1987, p.1.

7. Hoy, March 9, 1987. Lower figures were also reported in the house-by-house censuses conducted by the municipalities. In western Napo Province, the zone of primary impact from the disaster, the assessment team of CATEC/Catholic Relief Services also conducted a house-by-house census of need. CATEC (Corporación de Apoyo a la Tecnología y a la Comuniación) joined with Catholic Relief Services in designing and conducting an emergency assistance project in Napo Province. Both are voluntary relief organizations financed by contributions from Catholic parishioners and operating with an international mission of social service. However, there were no complete records of residents living in the area prior to the earthquake, leaving in doubt the actual number of persons killed in the disaster. Summary of Relief Program, Catholic Relief Services, Quito, Ecuador, June 15, 1987. Interview, Program Director, Catholic Relief Services, Quito, Ecuador, July 12, 1987.

8. Gen. Antonio Moral Moral, National Director of the Civil Defense, cited in Hoy, March 9, 1987, p.1. See also the United Nations ECLAC Report #87-04-406, op. cit., p. 1.

9. Hoy, March 10, 1987, p. 3A.

10. Interview, Padre, parish church, Borja, Ecuador, July 8, 1987; interview, President, Municipal Council, Baeza, Ecuador, July 8, 1987.

11. Hoy. Also, professional interviews with local government officials in Baeza, Borja, and El Chaco, July 9, 1987, and with the Project Director, Proyecto Emergencia in Napo Province, Catholic Relief Services/CATEC in Quito, July 12, 1987.

12. Hoy, March 10, 1987, p. 9. Interview, Field Representative, USAID/OFDA, Quito, Ecuador, June 28, 1987.

13. Interview, President of the Provincial Council, Imbabura, Ibarra, July 9, 1987. A census of damage to public buildings, infrastructure, and private homes was conducted by the Civil Defense councils of each canton in the province of Imbabura that suffered damage from the earthquakes. These cantonal reports were then forwarded to the Provincial Council of Civil Defense, Imbabura Province. The set of damage assessment reports was reviewed by the provincial government of Imbabura, and in turn, forwarded to the National Council of Civil Defense in Quito. These reports served as the basis for planning the reconstruction projects needed for the disaster zone. Essentially this same procedure was followed in all three zones of the disaster. Damage Assessment Reports, Provincial Government of Imbabura, Ibarra, Ecuador, March 19, 1987.

14. Professional observation, visit to Ibarra, Imbabura Province, Ecuador, July 9, 1987. See also news reports in Hoy, March 11, 1987, p.9A.

15. Professional observation and interviews, residents of Pedro Moncayo and President of the Municipal Council, Olmedo, June 19, 1987. See also news reports in Hoy, March 11, 1987.

16. Interview, President of the Municipal Council, Cayambe; July 9, 1987; interview, Secretary of the Municipal Council of Olmedo, July 9, 1987.

17. Interview, Padre, Misión Carmelita, Lago Agrio, Ecuador, June 29, 1987; interview, Director of the CEPE-Texaco Consortium, Quito, Ecuador, July 3, 1987. The CEPE-Texaco Consortium was established between the Government of Ecuador and Texaco Oil Company to manage oil production and shipment from a given location in eastern Napo Province.

18. Interview, Padre, Misión Carmelita, Lago Agrio, Ecuador, June 29, 1987; interview, Field Director for Indian Services, Catholic Relief Services, Quito, Ecuador, July 7, 1987.

19. The term "between" is used in a statistical sense to connote the type of variance that exists between member organizations of a given set, in contrast to the type of variance that exists "within" each member organization. In this analysis, the set of organizations includes all organizations that participated in the Ecuadorian disaster operations. Variance between organizations may be explained by distinctive characteristics or attributes of individual organizations. Variance within organizations is assumed to be distributed randomly. The total variance for the set of organizations is the sum of the between, or explained, variance and the within, or random, variance. This analysis is seeking to identify the types of characteristics that contribute to variance between organizations that participated in the Ecuadorian disaster operations.

20. Hoy, March 8, 1987, p.1; March 9, 1987, p.1; March 11, 1987, p.9A. These reports were confirmed in interviews during June-July 1987 with informed observers from both Ecuador and the United States, who had participated in disaster-assistance operations in March 1987. A survey of residents of the disaster zones also confirmed difficulties and delays in the distribution of disaster assistance. It is important to identify where the difficulties exist in the process, without making judgments as to cause, before the process can be redesigned for improved performance.

21. Interview, Coordinator, COEN (National Center of Emergency Operations), Quito, Ecuador, July 7, 1987. See also Secretaria General del Consejo de Seguridad Nacional, Dirección Nacional de Defensa Civil, Ley de Seguridad Nacional, 1987, p.1.

22. Interview, Comandante de Battallon de Selva, Lago Agrio, and Director of Civil Defense, Canton of Lago Agrio, Lago Agrio, Ecuador, June 30, 1987.

23. Interview, Coordinator, COEN, and director, National Council of the Civil Defense Authority, Quito, Ecuador, July 7, 1987.

24. Interview, National Director of Civil Defense, Quito, Ecuador, July 7, 1987.

25. Hoy, March 7, 1987, p. 6A.

26. Ibid.

27. Interview, Director of the national Emergency Committee and COEN; interview, national Director of Civil Defense, Quito, Ecuador, July 7, 1987.

28. Hoy, March 7, 1987, p. 6A.

29. Hoy, March 11, 1987, p. 1. Interview, U.S. Ambassador, Quito, Ecuador, July 6, 1987.

30. Interview, Comandadura de Defensa Civil, Province of Pichincha, Quito, Ecuador, June 17, 1987.

31. Pastor, Iglesia del Pacto Evangélica de Ecuador, Fondación Adelanto Comunitario Ecuatoriana, Quito, Ecuador, July 13, 1987.

32. Hoy, March 22, 1987, p.1. Interview, professor of sociology, Universidad Católica, Quito, Ecuador, June 22, 1987.

33. Professional observation of a *minga* in operation, Canton Cayambe, Ecuador, June 19, 1987.

34. Interview, Director, emergency assistance project in Napo Province, Baeza, Ecuador, July 8, 1987; interview, President, Municipal Council, Baeza, Ecuador, July 8, 1987.

35. Interview, President, Municipal Council, Cayambe, Ecuador, July 2, 1987.

36. This observation was made separately by several informed observers/ participants in the disaster response and recovery process. Assurances of professional confidentiality prevent citing the sources directly. The same observations are documented in news articles published in Hoy, March 7-31, 1987.

37. Interviews with directors of local and provincial Civil Defense councils and directors of disaster-assistance projects, Quito, Olmedo, Cayambe, Baeza, Borja, and Lago Agrio, Ecuador, June 19-July 14, 1987.

38. Interview, Vice-president, Municipal Council and Coordinator of the Municipal Civil Defense Committee, Baeza, Ecuador, July 8, 1987.

39. Interview, Director, Proyecto Emergencia, CATEC/CRS, Quito, Ecuador, July 12, 1987; interview, President, Municipal Council, Baeza, Ecuador, July 8,

1987; interview; Program Director, Peace Corps Ecuador, Quito, Ecuador, July 10, 1987.

40. Hoy, March 7-31, 1987; interviews, disaster assistance volunteers, Baeza, Ecuador, July 8-9, 1987; and professional observation, Baeza, Borja, El Chaco, and Tres Cruces, July 8-9, 1987.

41. Hoy, March 7-31, 1987.

42. Hoy, March 10, 1987, p.9A.

43. These figures were reported by Civil Defense Ecuador in their final report on the disaster.

44. Hoy, March 7-31, 1987. These figures were documented from several sources, including the Office of the Director of Housing for COEN, Quito, Ecuador, July 13, 1987.

45. Interviews, President of Municipal Council, Cayambe; President, Municipal Council, Olmedo; President, Provincial Council, Imbabura, July 2, 1987.

46. Hoy, March 11, 1987, p. 1. Interview, President, Municipal Council, Cayambe, July 2, 1987.

47. Interview, Field Representative, USAID/Office of Foreign Disaster Assistance, Quito, Ecuador, June 28, 1987.

48. Interview, President, Municipal Council, Cayambe, July 2, 1987.

49. Interview, Padre, Catholic Church, Cayambe, June 29, 1987. Interview, Madre Superiore, Catholic School, Cayambe, July 2, 1987.

50. Interview, Catholic volunteer worker, Olmedo, Ecuador, June 27, 1987.

51. Interview, President, Municipal Council, Cayambe, July 2, 1987.

52. Interview, Program Director, Catholic Relief Services, Quito, Ecuador, July 12, 1987.

53. Interview, Director, COEN, Quito, Ecuador, July 7, 1987. Hoy, March 21, 1987, p.9A.

54. Interview, Comandante de Battallon de Selva and Coordinator, Civil Defense Council, Canton of Lago Agrio, Lago Agrio, Ecuador, June 30, 1987. Hoy, March 8, 1987, p.8A.

55. Interview, Field Director for Indian Services, Catholic Relief Services, Quito, Ecuador, July 7, 1987.

56. Interview, Padre, Misión Carmelita, Lago Agrio, Ecuador, June 29, 1987.

57. Interview, Field Director, Catholic Relief Services, Quito, Ecuador, July 7, 1987.

58. Hoy, July 13, 1987. Interviews, informed observers from both Ecuador and the United States, Quito, Ecuador, June 16-July 15, 1987.

59. Interview, Assistant Director for Latin America, Office of Foreign Disaster Assistance, Washington, D.C., September 1, 1988.

60. Hoy, July 15, 1987. Interview, professor of public administration, Universidad Católica, Quito, Ecuador, June 22, 1987.

Appendix A

Disaster Management Organizations—General

NATIONAL

Public/Governmental

Centro de Operaciones Emergencia Nacional
  Comité de Emergencia
  All Ministries
  Defensa Civil
  Ejército
  Provincial, Cantonal, Parochial Mingas
  Universities:
    Escuela Politécnica Nacional
    Universidad General
Instituto Ecuatoriana de Minería
Instituto Ecuatoriano de Electricidad
Consortium Estatal Petrolera Ecuatoriana

Private/Voluntary

Cruz Roja Ecuatoriana
Catholic Relief Services
National "Crusade for
  Solidarity"
Utilities
Labor Unions
Public Employee
  Organizations
Universidad Católica
  de Quito

INTERNATIONAL

Public/Governmental

United Nations Disaster Relief Organization
Pan-American Health Organization
United Nations Development Program
National Embassies*
U.S. Mission:
  Office of Foreign Disaster Assistance
  Agency for International Development
  Peace Corps
  Military Group
  Embassy
American Community:
  Volunteers
  Businesses
  Texaco Oil Co.

Private/Voluntary

International Committee
  of the Red Cross
Catholic Relief Services
World Vision
Friends of the Americas
Save the Children

---

*Tended to work with host country on country-by-country basis. Efforts to coordinate actions among embassies were not very effective. It was very difficult and demanding to invent coordination on the spot. Establishing linkages between national and international organizations was too time-consuming.

Appendix B

International Organizations Involved in the March 1987 Disaster Operations

PUBLIC/GOVERNMENTAL

CAAP - Pacto Andino
CAF - Corporación Andina de Fomento
IDB - Interamerican Development Bank
IMF - International Monetary Fund
OAS - Organization of American States (OEA)
OPEC - Organization of Petroleum Exporting Countries (OPEP)
UNDRO - United Nations Disaster Relief Organization
UNESCO - United Nations Children's Fund
World Bank (IBD - International Bank of Development)

PRIVATE/VOLUNTARY

CRS - Catholic Relief Services
CATEC - Corporación de Apoyo a la Tecnología y a la Comunicación
Evangelical Brotherhood
Friends of the Americas
HCJB - Hoy Cristo Jésus Benedictus
ICRC - International Committee for the Red Cross
Save the Children
World Vision

NATIONAL GOVERNMENTS

Argentina
Belgium
Bolivia
Brazil
Canada
Chile
Colombia
Cuba
Federal Republic of Germany
France
German Democratic Republic
Great Britain
Holy See/Vatican
Italy
Japan
Malta
New Zealand

Norway
Paraguay
People's Republic of China
Peru
Spain
Switzerland
Union of Soviet Socialist Republics
United States:
    Agency for International Development
    Office of Foreign Disaster Assistance
    Army Corps of Engineers
    Embassy
    SOUTHCOM (military group)
    Peace Corps
    Geological Survey
Uruguay
Venezuela

Appendix C

Ecuadorian Organizations Involved in the March 1987 Disaster Operations

CENTRAL GOVERNMENT

Centro de Operaciones Emergencia Nacional (COEN)
    Fondo Nacional de Emergencia
    Comité de Emergencia: Chair, Gen. German Ruíz
    Members

Ministries:
    (1)   Financia
    (2)   Industria
    (3)   Salud
    (4)   Bienestar Social
    (5)   Energía y Minería
    (6)   Obras Publicas
    (7)   Transporte
    (8)   Agricultura y Ganadería
    (9)   Trabajo
    (10) Relaciones Extranjeros
    (11) Interiore

Defense:
    Defensa Civil
    Fuerza Armada: Ejército Ecuatoriano
                         Cuerpo de Ingenieros
                         Fuerza Aérea Ecuatoriana

Other:
  (1) Comandancia de la Policía
  (2) Instituto Ecuatoriano de Minería (INEMIN)
  (3) Instituto Ecuatoriano de Electricidad (INECEL)
  (4) Instituto Nacional de Estadística
  (5) Instituto Geofísico de Escuela Politécnica Nacional
  (6) Observatorio Astronómico de Quito
  (7) Instituto Ecuatoriano de Reforma Agrariana y Colonización
  (8) Fundación Nacional de Prevención (FONAPRE)
  (9) INHAMI
  (10) CONADE
  (11) Instituto Nacional de Niños y Familias

Consortium Estatal Petrolera Ecuatoriana (CEPE)
President of the National Congress
President of the Supreme Court

NATIONAL ORGANIZATIONS: PRIVATE/VOLUNTARY

Cruz Roja Ecuatoriana
Cruz Amarilla
Cruzada Nacional de Solidaridad (Crusade of Solidarity for the Victims of the Earthquake)
Comité de Coordinación y Control de Ayuda
Ham radio operators (RACES)
Federación Nacional de Comerciantes Minoritas
IBM of Ecuador
Empresa Eléctrica
CENTRAMECS (Centro Nacional de Trabajadores de Medios de Comunicación Social)
General Council of Chambers of Commerce
Episcopal Conference
Federación de Chóferes Profesionales
United Committee of the Strike – 24 March 87
Asociación de Damas de la Pequeña Industria (Women of Small Industry)
National Union of Ecuadorians
50 Women's Groups of Pichincha – 11 March 87
Voluntarios del Hospital Metropolitano
Comité de Provincias Asociadas de Ecuador
Asociación de Enfermas Incurables

Asociación de Mujeres de Ecuador (Women's Association of Ecuador)
Federación Nacional de Cooperativas de Ahorro y Crédito
Federación de Servidores Públicos de Pichincha
CONASEP (Confederación Nacional de Servidores Públicos)
Asociación de Empleados y Obreros
Frenta Unitaria de Trabajadores
CONAIE (Confederación de Nacionalidades Indígenas del Ecuador)
CTE (Confederación de Trabajadores de Ecuador)
Banco Nacional de Fomento
CONFENAIE (Confederación de Nacionalidades Indígenas de la Amazonia Ecuatoriana)
Union de Organizaciones Campesinos de San Pablo del Lago
Federación Unitaria de Organizaciones Sindicales
Federación de Cámaras de Pequeños Industriales
FOIN (Federación de Organizaciones Indígenas de Napo)
FECUNAE (Federación de Comunas de Napo Ecuatoriana)

LOCAL ORGANIZATIONS

Provincial, Cantonal, Parochial, and Barrial Councils of Government
  Provinces: Carchi, Imbabura, Napo, Pastaza, Pichincha
  Cantons: Cangahua, Lago Agrio, Quijos
Provincial, Cantonal, Parochial, and Barrial Councils of Civil Defense
Diocesan and Parochial Churches, overlapping religious organizations
Colegios and Schools
Ethnic associations, local chapters, and informal bonds
Communal *mingas* or cooperative work projects, organized to accomplish specific objectives
Community development projects, linked with national and international organizations
Community emergency committees